Wastewater Biology: The Habitats

A Special Publication

Prepared by **Wastewater Biology: The Habitats Task Force** of the **Water Environment Federation**

Michael H. Gerardi, *Chair*

Greg Anderson
Khalil Z. Atasi
Katherine Baker
Frank Barvenik
Mark T. Baumgardner
Stephen Bopple
Paul Bowen
Jerry T. Cheshuk
Steven C. Chiesa
Martha Anne Dow
Jerzy J. Ganczarczyk
Anthony Gaudy, Jr.
Joseph J. Gauthier
Roger Hlavek
Kenneth E. Kaszubowski
David J. Kinnear
Nancy Kinner
Nikki Korkatti
Timothy LaPara
Dennis R. Lindeke
Ken Mackenthun
Miryoussef Norouzian
Malcolm D. Peacock
Wesley O. Pipes
Lewis N. Powell
Raymond W. Regan
Sophie G. Simon
Edward V. Skonieczny
Gary L. Sober
Stuart J. Spiegel
Thomas L. Stokes, Jr.
Richard S. Talley
Allan R. Townshend
Robert A. Urban
T. Viraraghavan
Scott Walters
Randel L. West
Robert Wichser
Melvin C. Zimmerman

Under the Direction of the
Municipal Subcommittee of the **Technical Practice Committee**

2000

Water Environment Federation
601 Wythe Street
Alexandria, VA 22314-1994 USA

IMPORTANT NOTICE

The contents of this publication are not intended to be a standard of the Water Environment Federation (WEF) and are not intended for use as a reference in purchase specifications, contracts, regulations, statutes, or any other legal document.

No reference made in this publication to any specific method, product, process, or service constitutes or implies an endorsement, recommendation, or warranty thereof by WEF.

WEF makes no representation or warranty of any kind, whether expressed or implied, concerning the accuracy, product, or process discussed in this publication and assumes no liability.

Anyone using this information assumes all liability arising from such use, including but not limited to infringement of any patent or patents.

ABSTRACT

This special publication reviews the natural and wastewater environments or *habitats* of microscopic and macroscopic organisms associated with wastewater treatment systems. Commonly occurring organisms and their basic roles in wastewater treatment are identified, and their habitats and the way these habitats and operational conditions interact are described. Chapter 1 introduces the primary microscopic and macroscopic groups of organisms associated with wastewater treatment processes and the types of habitats in which they are commonly found. Chapter 2 provides a simplified presentation of the interactions of significant populations of organisms (e.g., *bacteria* and *protozoa*) within wastewater treatment processes and the way these interactions affect the *degradation* of wastes. Chapter 3 reviews *floc* formation in the *activated sludge* process, organisms responsible for floc development, and operational factors responsible for the interruption of proper floc formation. Chapter 4 describes the different environmental zones—*aerobic*, *microaerophilic*, and *anaerobic*—found in fixed-growth processes, the dominant organisms associated with each zone, and the way these zones and their organisms degrade waste. Chapter 5 reviews different environmental zones within wastewater stabilization lagoons, their respective dominant organisms, and the way lagoons degrade wastes. Chapter 6 reviews the uses of *wetlands* for wastewater treatment. Finally, Chapter 7 describes the process of *composting* from a biological perspective, the composition of compost, and organisms associated with its development.

Library of Congress Cataloging-in-Publication Data

Wastewater biology: the habitats: a special publication/prepared by Wastewater Biology—the Habitats Task Force of the Water Environment Federation; under the direction of the Municipal Subcommittee of the Technical Practice Committee.

p. cm.

Includes bibliographical references and index.

ISBN 1-57278-163-7

1. Sewage—Microbiology. 2. Microbial ecology. 3. Sewage—Purification—Biological treatment. I. Water Environment Federation. Wastewater Biology: the Habitats Task force. II. Water Environment Federation. Municipal Subcommittee.

TD736.W35 2000
628.1'62—dc21 00-063335

ISBN 1-57278-163-7
Printed in the USA

Water Environment Federation

Founded in 1928, the Water Environment Federation is a not-for-profit technical and educational organization. Its goal is to preserve and enhance the global water environment. Federation members number more than 38,000 water quality professionals and specialists from around the world, including engineers, scientists, government officials, utility and industrial managers and operators, academics, educators and students, equipment manufacturers and distributors, and other environmental specialists.

For information on membership, publications, and conferences, contact

Water Environment Federation
601 Wythe Street
Alexandria, VA 22314-1994 USA
703-684-2400
http://www.wef.org

Special Publications of the Water Environment Federation

The WEF Technical Practice Committee (formerly the Committee on Sewage and Industrial Wastes Practice of the Federation of Sewage and Industrial Wastes Associations) was created by the Federation Board of Control on October 11, 1941. The primary function of the Committee is to originate and produce, through appropriate subcommittees, special publications dealing with technical aspects of the broad interests of the Federation. These publications are intended to provide background information through a review of technical practices and detailed procedures that research and experience have shown to be functional and practical.

Contents

List of Tables

List of Figures

Preface

This publication is the third in a series of publications on wastewater biology developed by the Water Environment Federation. The first two texts in the series, *Wastewater Biology: The Microlife* and *Wastewater Biology: The Life Processes*, were originally published in 1990 and 1994, respectively. An update to *Wastewater Biology: The Microlife* is expected in 2001.

The newest addition to this series of texts is *Wastewater Biology: The Habitats*. This text reviews the natural and wastewater environments or habitats of the microscopic and macroscopic organisms associated with wastewater treatment systems. Written with a minimum of technical jargon, the text provides not only an identification of commonly occurring organisms and their basic roles in wastewater treatment, but also a description of their habitats and the way these habitats and operational conditions interact. *Wastewater Biology: The Habitats* can be used as an operational and troubleshooting text, a reference text, or an educational resource.

This publication was produced under the direction of Michael H. Gerardi, *Chair*. Principal authors and chapters for which they were responsible are

Martha Anne Dow (5)
Jerzy J. Ganczarczyk (3)
Joseph J. Gauthier (7)
Michael H. Gerardi (1)
David J. Kinnear (7)
Nancy Kinner (4)
Raymond W. Regan (2)
Melvin C. Zimmerman (6)

Authors' and reviewers' efforts were supported by the following organizations:

Bio Systems Corporation, Roscoe, Illinois
Blue Heron Environmental Technology, Athens, Ontario, Canada
Brown and Caldwell, Salt Lake City, Utah
City of Yorkton Water Pollution Control, Yorkton, Saskatchewan, Canada
Cytec Industries, Stamford, Connecticut
Dover Township, York, Pennsylvania
Drexel University, Philadelphia, Pennsylvania
EMA, Inc., Harrisburg, Pennsylvania
Environmental Research & Design, Salt Lake City, Utah
Fort Worth Water Department, Fort Worth, Texas
Gills, Guard and Johnson, Inc., Willoughby, Ohio
Lycoming College, Williamsport, Pennsylvania
Kenneth M. Mackenthun, Inc., Vienna, Virginia
McNamee, Porter & Seeley, Inc., Detroit, Michigan

Metcalf & Eddy, Inc., Atlanta, Georgia
Metro Waste Control Commission, Hastings, Minnesota
O'Brien & Gere Engineers, Inc., Syracuse, New York
Oregon Institute of Technology, Klamath Falls, Oregon
Parkhill, Smith & Cooper, Inc., Lubbock, Texas
Pennsylvania Department of Environmental Resources, Harrisburg, Pennsylvania
Pennsylvania State University, University Park, Pennsylvania
R.E. Warner & Associates, Westlake, Ohio
Richmond Department of Public Utilities, Richmond, Virginia
Stokes Environmental Associates, Ltd., Norfolk, Virginia
Santa Clara University, Santa Clara, California
Sigma Environmental, Oak Creek, Wisconsin
Uni-Tec Consulting Engineers, State College, Pennsylvania
University of Alabama, Birmingham, Alabama
University of New Hampshire, Durham, New Hampshire
University of Regina, Regina, Saskatchewan, Canada
Veeco, Covington, Louisiana
Village of Minerva Park, Columbus, Ohio

Chapter 1
Introduction

OVERVIEW

Wastewater Biology: The Habitats is an overview of the biological, chemical, and physical living areas or *habitats* occupied by the microscopic and macroscopic organisms involved in wastewater *stabilization*. The habitats consist of numerous *microbial communities* such as *floc* particles (suspended growth), *biofilm* (fixed growth), and natural *wetlands*, which contain a large diversity of organisms, including *bacteria*, *protozoa*, *rotifers*, *free-living nematodes*, *bristleworms*, *flatworms*, *waterbears*, and immature and adult insects (Table 1.1).

This publication reviews the interrelationships among the organisms and their habitats (or environments). An understanding of these interrelationships is of value to operators in their attempts to successfully regulate and troubleshoot wastewater treatment processes. This information will also be of value to other wastewater professionals, including environmental engineers, chemists, microbiologists, and educators, in addressing wastewater treatment processes from not only a *quantitative* approach but also a *qualitative* focus.

Wastewater Biology: The Habitats is the third text in a series of specialty publications on wastewater biology. The first text in the series, *Wastewater Biology: The Microlife* (WPCF, 1990), presented the description, ecology, and beneficial and detrimental roles of the microscopic life forms—the microlife—found in wastewater treatment processes, while the second text, *Wastewater Biology: The Life Processes* (WEF, 1994), presented the biological activities—the life processes—performed by the microlife in stabilizing wastewater and residual solids.

This text contains numerous illustrations and a minimum of technical jargon so that it can be easily read and interpreted by individuals with

Table 1.1 Commonly observed organisms in wastewater habitats.

Actinomycetes, e.g., *Nocardia amarae*
Algae
Bacteria
Autotrophs or chemolithoautotrophs, nitrifying bacteria
Heterotrophs or organotrophs, saprophytic bacteria
Filamentous bacteria, e.g., *Sphaerotilus natans*
Floc-forming bacteria, e.g., *Escherichia coli*
Bristleworms
Crustaceans
Flatworms
Fungi
Insects
Immature, e.g., midge larvae, *Chironomus* sp.
Adult, e.g., water scavenger beetle, *Hydrophilus* sp.
Nematodes
Protozoa
Amoebae, e.g., *Arcella* sp.
Flagellates, e.g., *Paranema* sp.
Free-swimming ciliates, e.g., *Paramecium* sp.
Crawling ciliates, e.g., *Aspidisca* sp.
Stalked ciliates, e.g., *Vorticella* sp.
Rotifers
Waterbears

different professional backgrounds. The text provides appropriate background information in a chapter on microbial ecology (Chapter 2) and discusses the common organism habitats encountered during treatment processes: floc particles or suspended growth (Chapter 3, Activated Sludge Floc), biofilms (Chapter 4, Fixed-Growth Processes), treatment lagoons (Chapter 5, Wastewater Stabilization Lagoons), natural and artificial wetlands (Chapter 6, Wetlands), and *composting* operations (Chapter 7, Composting).

All chapters within this publication are concerned with several basic interrelationships as shown in Figure 1.1 and the transfer of energy in wastewater treatment systems from *biochemical oxygen demand (BOD)* through a *food web* of unicellular to multicellular organisms as shown in Figure 1.2.

INTERRELATIONSHIPS

Any artificial or constructed treatment system consists of three basic components: the operator, the environment, and the organism (Figure 1.1). The action of one component has either a direct or indirect effect on another component.

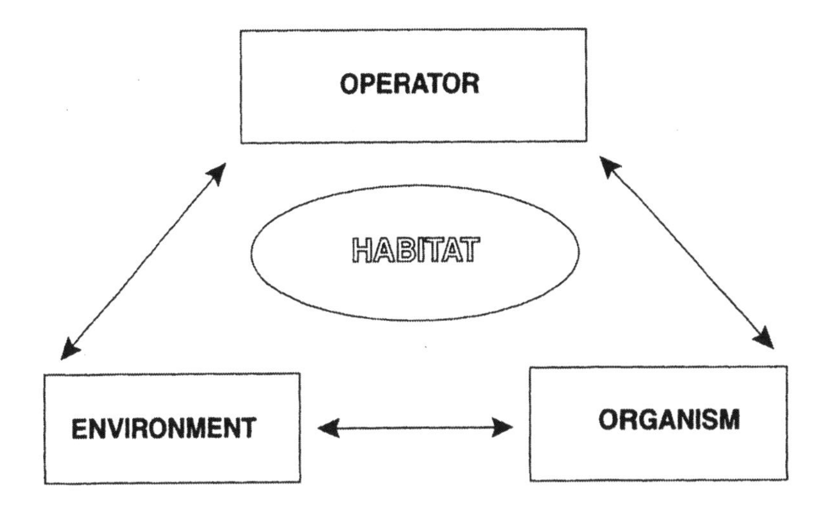

Figure 1.1 Interrelationships within a wastewater treatment process.

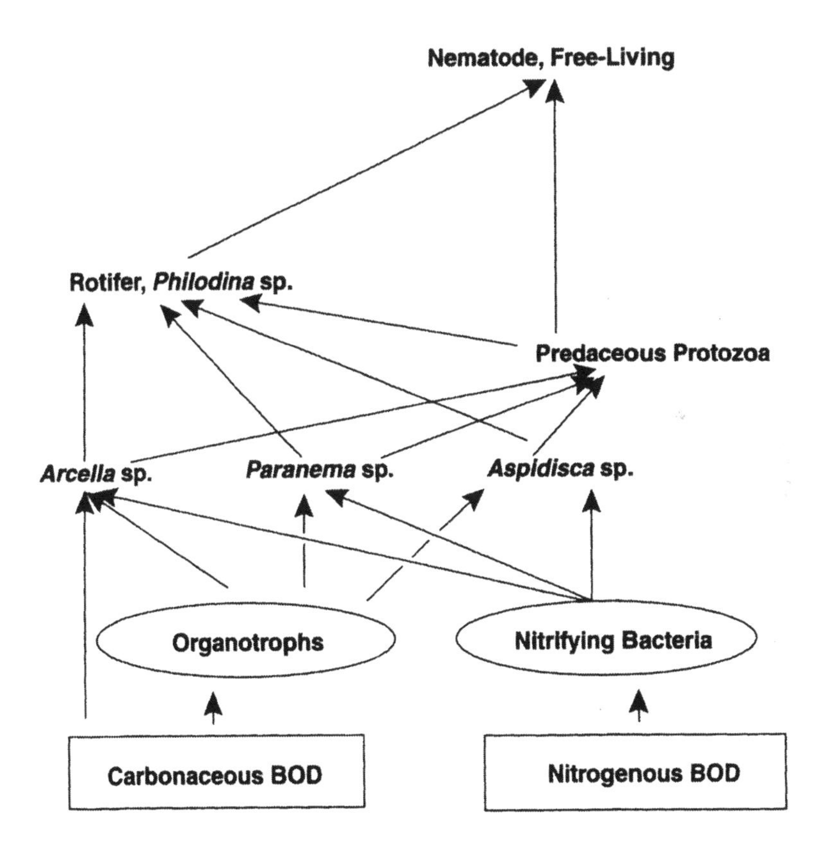

Figure 1.2 A food web within a wastewater treatment process.

For example, changes in wastewater strength and composition (environment) determine the relative abundance, diversity, and *dominance* of the life forms within the habitat and the efficiency of the treatment process. If the strength and composition of the wastewater should suddenly change, for example, an accidental discharge of excessive oils and greases to an *activated sludge* process, rapid growth or *proliferation* of foam-producing *filamentous organisms* such as *Nocardia* spp. may occur, resulting in *settleability* problems and loss of solids. This condition may cause the operator to initiate *chlorination* of the *return activated sludge* (RAS), which, in turn, may adversely affect nitrifying bacteria within the activated sludge, resulting in loss of *nitrification*.

Another example would be the production of an *anoxic* environment in the secondary *clarifiers*, resulting in *denitrification* and, again, settleability problems and loss of solids. This condition may cause the operator to increase the RAS rate to overcome denitrification, that is, the effects of denitrifying bacteria.

Each of these examples is an illustration of changes occurring in the relative abundance, diversity, activity, and dominance of the life forms within the floc particle or habitat and the interrelationships between the operator, the environment, and the organism.

FOOD WEB

The *food web* found in wastewater treatment processes consists of the energy found in the chemical bonds of compounds within *carbonaceous BOD* and *nitrogenous BOD* at the bottom of the web, then the unicellular organisms (primarily bacteria), and then the large diversity of unicellular and multicellular organisms, primarily protozoa, rotifers, and free-living nematodes (Figure 1.2) at the top of the web. The more successful the operator is in transferring the energy throughout the food web, the more abundant, active, and diverse the life forms are within the habitat. The more abundant, active, and diverse the life forms are within the habitat, the more successful the operator is in transferring energy from the chemical bonds of carbonaceous BOD and nitrogenous BOD and the more successful the operator is in operating the treatment process.

By understanding how the organisms change their environment (for example, nitrifying bacteria destroy *alkalinity* in the treatment process) and how the environment favors or disfavors the activity of organisms (for example, low *pH* and alkalinity inhibits the activity of nitrifying bacteria), operators are better able to troubleshoot treatment plant processes (for example, if pH and alkalinity are dropping, why they are dropping and what can be done to maintain a desired environment). This text should reinforce that operators are critical to the treatment process and their actions or inactions have a profound effect on the habitats of the organisms and the success or failure of the process.

References

Water Environment Federation (1994) *Wastewater Biology: The Life Processes.* Special Publication, Alexandria, Va.

Water Pollution Control Federation (1990) *Wastewater Biology: The Microlife.* Special Publication, Alexandria, Va.

Chapter 2
Microbial Ecology

INTRODUCTION

Biological wastewater treatment relies on a large diversity of organisms to remove contaminants from wastewater. These organisms are a *community* of many different *microbial species* (i.e., a mixed population), and the organisms interact with one another and their environment. In general, the nature of the community influences efficiency and effectiveness of the treatment process.

BACKGROUND

DEFINITION OF MICROBIAL ECOLOGY. *Microbial ecology* may be defined as the study of the interactions of organisms with their physical environment and with one another. During the study of microbial ecology, how an organism affects, and is affected by, its environment and how these interactions determine the kinds and numbers of organisms found in a particular place and time are determined (Atlas and Bartha, 1987).

IMPORTANCE OF MICROBIAL COMMUNITIES. Each type of organism within the community can be critical to treatment process success. For example, *gravity separation* of *activated sludge biomass* from treated *effluent* is critical for successful operation of the overall plant. If *filamentous organisms* are a substantial portion of a microbial community, the plant may fail altogether or efficiency may be reduced because these microbes settle slowly, causing secondary *clarifier* efficiency to be low. Likewise, if an unusual pollutant is present in the waste, it may not be removed unless a group of organisms with the ability to remove that pollutant is present in sufficient numbers.

Microbial composition and the role (*niche*) of the members of the community are important to this discussion of microbial-community organization. The term *dominance* is frequently used to describe the community. A *species*, or a group of species, may be considered *quantitatively* (or numerically) dominant if their numbers are larger than another species. Dominance may also be based on the mass (or amount) of the particular species of interest or whether the relative growth rate of a species is much greater than another. Likewise, dominance may be based on one species imparting a certain trait on the community. For example, the presence of relatively few filamentous organisms may be enough to cause *bulking*. Thus, dominance may be based on a number of factors, and the basis must be stated when dominance of a microbial community is discussed.

ORGANIZATION OF MICROBIAL COMMUNITIES

TROPHIC LEVELS. *Trophic levels*, or *food chains* and *food webs*, are commonly used to describe the organization of communities. Food chains and food webs trace the flow of energy through the system. *Primary producers* convert the energy of the sun to carbon-based energy. In land communities, primary producers are the green plants. Thus, land-based wastewater treatment processes require primary producers. Shallow stabilization lagoons, for example, may include *algae* as primary producers.

Wastewater (*biochemical oxygen demand [BOD]*, *carbonaceous BOD*, and *nitrogenous BOD*) is the principal source of energy in many wastewater treatment processes so that primary producers are not needed. In wastewater treatment systems, the lowest trophic level is occupied by the primary *saprophytes* (organisms that consume nonliving organic matter) (Ray, 1995). Saprophytes, including various bacterial species (*aerobes* and *facultative anaerobes*), are considered the most important organisms of the microlife because they are responsible for conversion of the *soluble* and *particulate organic matter* (measured as BOD) to carbon dioxide, new cells, and other end products. In anaerobic *digesters*, acid-forming bacteria are the primary saprophytes.

Some of the products from *metabolism* of primary saprophytes can be used by other organisms as sources of carbon and energy. The organisms that can consume products of other organisms are called secondary saprophytes. In the anaerobic digestion process, the activity of secondary bacterial saprophytes (*methane formers*) combines hydrogen and carbon dioxide to form methane.

At the top of the food chain or food web in many wastewater treatment processes are the *predators*. Predators use living *biomass* as their carbon and energy sources. The most common predators in activated sludge systems are the stalk ciliated *protozoa*. These protozoa remove dispersed bacteria that would otherwise end up in the secondary *effluent*. The species, their activity and structure, and number of protozoa may be used as an indicator of the health of an activated sludge process (WPCF, 1990). Other higher life forms may be present in the activated sludge *microbial community* and serve as further indicators of process stability. For example, a limited number of *rotifers* are considered to be desirable, while excess numbers may indicate a condition associated with a *turbid* effluent (McKinney, 1962).

Other predators, including snails, flies, and other *macroorganisms*, may also be present in *biofilm* processes. In general, these macroorganisms do not contribute to the overall process of removing *organic* materials from wastewater and may cause periodic nuisance conditions, for example, with biomass settling in secondary clarifiers and clogging pumps and nozzles in activated sludge processes. A more serious problem may be the infestation of biofilm systems by worms or insect *larvae* because they consume large amounts of the biofilm itself (see Chapter 4).

THE NICHE. The niche is another useful means of describing the organization of microbial communities. The niche of an organism is its role performed in the microbial community (Krebs, 1972). It is defined by the compounds used as energy and carbon sources, the physical space in which an organism resides (for example, on the interior of an activated sludge *floc* or the biofilm of a *trickling filter*), and other factors.

A concept of critical importance to the niche concept is the *competitive exclusion principle*, which states that only one species can occupy any given niche for an extended period of time (Krebs, 1972). In other words, if two species use the same environmental resources (such as organic *substrate* or food), by competition only one will survive. Competition can be thought of as the survival of the fittest.

DIVERSITY OF NICHES AND MICROORGANISMS. Competitive exclusion theoretically limits the number of species of organisms and niches that may be present in a wastewater treatment system. Thus, a system with a *homogeneous* environment (one with little diversity) and very few organic compounds in the wastewater may have a limited number of microbial species. This has been found to be the case with some industrial wastewater treatment systems where a limited microbial population develops in response to the specific substrates present.

The physical environment provided in the design of a wastewater treatment process may add to the diversity of niches. Biofilm processes are a prime example of this physical diversity. The environment in a trickling filter, for example, varies from the aerobic surface to the *anaerobic* inner layer of the biofilm. Activated sludge environments also have physical diversity in which the organisms inhabiting the interior of an activated sludge floc particle may be in an oxygen-deficient environment while those on the exterior of the floc are exposed to *dissolved oxygen*.

An additional degree of physical diversity is added to activated sludge processes by settling and recirculation of the sludge. The flow of activated sludge organisms through a series of oxygenated and *anoxic* (oxygen-deficient) zones in the biological reactor (i.e., *aeration* tank) and the final clarifier sludge blanket, respectively, will add to the number of niches possible in the system. For example, a niche may be available to denitrifiers in nitrifying activated sludge systems because these organisms can metabolize carbonaceous BOD while they are in the anoxic sludge blanket of the secondary clarifier.

The number of niches in a wastewater treatment process may also be increased by the cyclic loading pattern typical for many treatment plants. Daily, weekly, and seasonal variations in loading may increase the chances for survival of some microlife.

The large amount of diversity in municipal wastewater treatment processes results in a large number of possible niches and, in turn, a large diversity of microbial species present. Studies of the bacteria present in municipal wastewater and activated sludge have found more than 400 species present, although many fewer species may dominate the microlife (WPCF, 1990, and Lighthart and Loew, 1972). Thus, the microbial communities present in typical wastewater treatment processes are very diverse and complex. Control of such mixed populations requires the operator to have an understanding of *population dynamics*.

POPULATION DYNAMICS

The individual species in the microbial community of a treatment process must compete to establish themselves in a niche and, thus, survive in that process. This pattern of competition and survival is referred to as *population dynamics*. Major factors in population dynamics (or *succession*) are competition for the same substrate (food) and nutrients, *predator–prey relationships*, *symbiotic* associations, and *growth rate* (Ray, 1995, and McKinney, 1962).

COMPETITION. In terms of *competition* among species, the following guidelines have been identified (McKinney, 1962). Two bacterial species with a comparable ability to *metabolize* a given substrate at the same rate will survive in equal masses (equal weights of microorganisms) rather than in equal numbers of microorganisms. Because many bacteria in wastewater treatment processes are approximately the same size, survival is related to the ability of the species to establish a niche. One important niche is the ability to metabolize a given type of substrate, measured as BOD. *Pseudomonas* spp. metabolize a wide range of substrates and survive in almost every wastewater microbial community. *Alcaligenes* spp. and *Flavobacterium* spp. metabolize primarily *proteins*.

Competition for the same substrate occurs with *fungi*, algae, protozoa, and bacteria. The smaller bacteria metabolize *soluble* substrate at an increased rate, followed by the fungi and then the protozoa at lower rates. When wastewater containing an increased substrate concentration is treated by an activated sludge process, all of these organisms grow, to some degree, with the bacteria predominating.

When the soluble organic concentration is decreased because a high *mixed liquor suspended solids (MLSS)* level is maintained in a biological reactor operated at a conventional *hydraulic retention time* of 6 to 8 hours, the freely suspended (nonflocculated) bacteria may decrease from (a) competition with greater populations of flocculated bacteria and (b) predation by protozoa. Bacteria are typically dominant over fungi and protozoa under optimum environmental conditions. A shift to fungi or filamentous organisms indicates adverse conditions that require operational changes. For example, at *pH* levels between 6.5 and 8.5, bacteria predominate over fungi. Below pH 6.5, the fungi are able to compete more successfully with bacteria. With decreasing pH, fungi predominate.

For a treatment process receiving only an intermittent loading of an industrial waste (for example, a discharge one day each month), the primary predominant bacteria develop initially to establish a niche. When the industrial waste has been removed, these microbes *lyse* (break up) and die, releasing proteins. *Flavobacterium* spp. and *Alcaligenes* spp. may have secondary predominance. If the time period between loadings is excessive, reduced treatment efficiency may result because the primary microbial population would have died off and been replaced by the secondary microbes, which are less suited for metabolizing the original industrial waste (McKinney, 1962).

PREDATOR–PREY RELATIONSHIPS. A major competition for substrate in wastewater treatment microbial communities involves bacteria (*prey*) and higher forms (predator). Bacteria competitively use soluble BOD as substrate. The higher microlife, primarily protozoa, use dispersed bacterial cells because of their inability to compete successfully with bacteria for soluble BOD.

SYMBIOTIC ASSOCIATIONS. Beneficial relationships among microbial members of a community (symbiosis) exist for activated sludge, anaerobic digestion, *stabilization* lagoons, and biofilm processes. Some of these relationships may be considered to be *commensal*, in that one organism benefits more than the other.

A number of possible symbiotic relationships that may be important in activated sludge processes have been reported, including

- *Flocculation* of particulate organic matter by one organism, which is then consumed by other organisms (Pipes, 1966).
- Production of a metabolic product by one organism, which is then metabolized by others (Pipes, 1966). An example is *oxidation* of ammonia to nitrite by *Nitrosomonas,* which is further oxidized to nitrate by *Nitrobacter*.
- *Degradation* of *cellulose* or paper products to yield glucose, which is then used as a substrate by other bacteria (Gerardi and Grimm, 1987).
- *Assimilation* of organic matter by the MLSS as a whole. Mixed cultures are usually more effective in reducing wastewater BOD and forming microbial floc than are pure cultures (Pipes, 1966).

Anaerobic digesters illustrate another symbiotic relationship in which the activity of one organism alters a feature of the physical environment, making it suitable for the growth of a second organism. For example, facultative anaerobes may consume all of the oxygen present in a tank containing organic wastewater biomass, thereby allowing growth of *obligate anaerobes* (no oxygen) (Ray, 1995).

Symbiosis involving bacteria and algae are found in the operation of shallow stabilization lagoons under sunlight conditions. Although they do not compete for substrate, their activities are interrelated. In the lagoon, aerobes consume organic matter from wastewater to yield several products, including new cells, carbon dioxide, water, ammonia, and other *inorganics*, in the presence of sunlight. Algae use carbon dioxide and other inorganic nutrients formed by the bacteria to create new algae cells. Oxygen, which is then used by the aerobes, is released into the lagoon water as a result of *photosynthesis*. The cycle continues during the daylight period. The association between these two life forms, however, is not always symbiotic. At night, both types of microbes respire, thereby competing for oxygen.

Insect larvae may have a beneficial effect on trickling filter (biofilm) slime that may be considered to be a symbiotic relationship (Gerardi and Grimm, 1987). Insects grazing on the microbial slime layers assist in maintaining the porosity of the filter for the penetration of oxygen needed to maintain aerobic conditions and in minimizing ponding. Consumption of the slime-forming organisms by the insects stimulates microbial growth, with a corresponding degree of substrate removal from the wastewater.

GROWTH RATES. The growth rate of each individual organism is a reflection of its presence in the microlife of the community. Commonly, wasting rates for conventional activated sludge processes provide a *solids retention time (SRT)* of 3 to 10 days. Some forms of the microlife, such as *nematodes* and nitrifying bacteria, are removed from the system at shorter SRTs. Because of their slower growth rates, these organisms cannot compete with the more rapidly growing *heterotrophic bacteria*. Also, nitrifier growth is limited by seasonal temperature variations, as discussed in the following section.

PROCESS OF ADAPTATION

IMPORTANCE. *Adaptation* can be defined as a change in a microbial community resulting from a change in the environment in which a community lives (Spain and VanVeld, 1983). This term is often used interchangeably with the term *acclimation.* Both of these terms have been used to describe the changes that occur when the microbial community is exposed to a new and unusual substrate or high concentrations of a usual substrate. Also, adaptation can refer to changes in the community resulting from changes in the physical environment, such as changes in temperature.

The ability of a microbial community to adapt to a new environmental condition can often result in success or failure of the treatment system. For example, in a municipal activated sludge process, the microbial community may be required to adapt to new organic compounds introduced by industrial dischargers (Melcer and Bedford, 1986). Likewise, an industrial wastewater treatment facility is sometimes faced with the need for its microbial community to adapt to changes occurring in the production facility. Adaptation can also occur in response to higher concentrations of compounds than the microbial community is accustomed to handling. For example, anaerobic treatment processes can adapt to high concentrations of ammonium (Ripley et al., 1984). Without a suitable acclimation period, however, the ammonium would be *toxic* to the microbial community.

Knowledge of how adaptation occurs is of utmost importance in situations when the treatment process is receiving toxic wastewater from an industrial source, especially in light of implementation of *toxicity* testing of effluent. The operator must be aware of how the treatment process may adapt to a given toxic compound to ensure its removal from the discharge.

MECHANISMS. The potential traits of a microbial community are defined by the species of organisms that may become significant members of the community, changes in the *genes* of the organisms that may take place through evolution, and *enzymes* that the organisms may synthesize. Microbial communities typically function with a very small number of the traits that are potentially available. Adaptation naturally occurs because it takes time for new organisms to become members of the community, mutations to occur, and enzymes to become synthesized in response to changes in the substrates present.

The mechanisms of adaptation revolve around the same factors that define the potential traits of a community. Thus, two primary mechanisms for adaptation are thought to exist (Spain et al., 1986): synthesis of specific enzymes not present in the population before the change in the environment and proliferation of specific species in the community that were not present in significant numbers before the change in the environment. Detailed discussion of these adaptation mechanisms is beyond the purpose of this publication. The following discussion provides a brief definition of key terms and a few examples to illustrate the importance of microbial adaptation in the operation of wastewater treatment facilities.

The processes involved in *enrichment* (or increasing the numbers) of organisms in a community can be deduced from what has been discussed in this chapter. For example, when a new compound (or substrate) enters the environment, an additional niche is potentially created. The niche may be filled by organisms entering the system, those already in the system, or others with changed genetic capabilities. If more than one organism tries to fill the niche created by the new substrate, competition for this substrate will result. The organisms surviving the competition will be those that can use the substrate most efficiently at the concentration at which the substrate is present. These organisms will gradually increase in number as the substrate is metabolized. When its numbers are sufficiently increased, an observable rate of degradation occurs.

An example of an enrichment that commonly occurs in municipal activated sludge processes is development of a nitrifying population, as discussed previously. This enrichment is often the result of a change in the environment of the system, such as an increase in temperature during the warmer period of the year. As temperature increases, the few nitrifiers in the system grow at an increased rate. If this rate is higher than the rate at which they are removed by solids wasting, the number of nitrifiers increases with observable *nitrification* occurring. During the cooler seasons of the year, the nitrifier growth rate decreases, and nitrification of the wastewater is more difficult to achieve.

The time required for enrichment of a new organism in the community is very difficult to predict. Among other things, it will depend on the number of organisms initially in the system and the number entering the system. If a toxic compound is added to the system, its concentration will influence the time required for enrichment of a new organism that can degrade the compound. If the concentration of the compound is too high, a toxic condition to all organisms in the community may be created. If the concentration is low, only a few of the desired organisms may grow and become established.

Comparing the orders of magnitude, enrichment may take days or months. For activated sludge processes, an adjustment period equal to three SRTs is often reported as being required for establishing a new microbial community (McKinney, 1962). This is a very general guideline that is based on hydraulic removal of an *inert* compound from an activated sludge process. It is likely that the complete establishment of a new microbial community may require a longer period.

AIDING THE PROCESS. Because understanding of the process of adaptation is incomplete, the ways in which operators can assist the process are limited. However, there are a number of actions that can be taken. The system can be seeded with the required organism to speed up the enrichment process. For example, enrichment of nitrifiers may be aided by simply seeding the system with activated sludge from a facility that is nitrifying. A similar approach may be taken when a new industrial waste is to be treated at a given facility. Using sludge from a system treating a similar waste as seed for the facility may reduce the time required for adaptation. *Bioaugmentation* with a purchased culture may also stimulate enrichment as long as the organisms added can survive in the process.

Rather than occasional or periodic discharge of a difficult-to-degrade substance to the treatment plant, the industry generating the pollutant can be required to provide a controlled continuous discharge (Melcer and Bedford, 1986). Continuous discharge of the pollutant will maintain an adapted microbial community while, on the other hand, an occasional discharge may cause the adaptation to be lost.

The discharger of a new potentially toxic pollutant can be required to add the pollutant to the wastewater initially at a decreased rate and gradually increase the level. Doing so should allow time for adaptation of organisms capable of degrading the toxic compound without adversely affecting other members of the microbial community. Alternatively, the discharge should be submitted to *bench-scale testing* for *treatability studies* and for estimating the appropriate process conditions and time needed for adjustment of the micro-life to the new conditions. If adaptation is successful, little or none of the toxic compound should pass through the process untreated.

The two most critical design features affecting establishment of a particular microbial population in the activated sludge process are the ability to transfer enough oxygen and the ability to waste solids from the system. Typically, the facilities experiencing problems are overloaded and cannot transfer enough oxygen at peak loading. Wasting systems are usually designed to operate under no-stress conditions. When stress occurs, the microbial population often becomes structurally unstable. As a result, the return sludge becomes thinner as thickening decreases in the secondary clarifier.

SUMMARY

Understanding the nature of the microbial community in a wastewater treatment plant is a very complex task. A complete understanding, although desirable, may not always be critical to successful operation of a particular treatment process. Under typical conditions, it may be sufficient for the operator to consider the microbial community as one microlife mass that responds as a whole to environmental changes and produces a relatively stable response. For example, this concept may be suitable for an adequately sized, municipal activated sludge process required to meet only conventional standards (that is, 30 mg/L BOD or suspended solids). In such a process, the activated sludge microbes respond as one unit to typical variations in loading while producing an acceptable effluent quality. However, for the unusual conditions, a more in-depth understanding of microlife responses may be needed to maintain effluent standards. For these conditions, the information in this chapter will be helpful.

References

Atlas, R.M., and Bartha, R. (1987) *Microbial Ecology: Fundamentals and Applications*. Benjamin/Cummings Publishing, Menlo Park, Calif.

Gerardi, M.H., and Grimm, J.K. (1987) Insects Associated with Wastewater Treatment: Their Role and Control. *Water Pollut. Control Assoc. Pa. Magazine*, 219.

Krebs, C.J. (1972) *Ecology, the Experimental Analysis of Distribution and Abundance*. Harper and Row, New York.

Lighthart, B., and Loew, G.A. (1972) Identification Key for Bacterial Clusters from an Activated Sludge Plant. *J. Water Pollut. Control Fed.*, **44**, 2078.

McKinney, R.E. (1962) *Microbiology for Sanitary Engineers*. McGraw-Hill, New York.

Melcer, H., and Bedford, W.K. (1986) Removal of PCP in Municipal Activated Sludge Systems. Presented at Water Pollut. Control Fed. 59th Annu. Conf., Los Angeles.

Pipes, W.O. (1966) The Ecology Approach to the Study of Activated Sludge. *Advances Appl. Microbiol.*, **8**, 77.

Ray, B.T. (1995) Microbiology and Microbial Growth. In *Environmental Engineering*. PWS Publishing, Boston, Mass.

Ripley, L.E.; Mohr-Kinet, N.; Boyle, W.C.; and Converse, J.C. (1984) The Effects of Ammonia Nitrogen on the Anaerobic Digestion of Poultry Manure. *Proc. 39th Ind. Waste Conf.*, Purdue Univ., West Lafayette, Ind.

Spain, J.C., and VanVeld, P.A. (1983) Adaptation of Natural Microbial Communities to Degradation of Xenobiotic Compounds: Effects of Concentration, Exposure Time, Inoculum, and Chemical Structure. *Appl. Environ. Microbiol.*, **45**, 428.

Spain, J.C.; Pritchard, P.H.; and Bourquin, A.W. (1986) Effects of Adaptation on Biodegradation Rates in Sediment/Water Cores from Estuarine and Freshwater Environments. *Appl. Environ. Microbiol.*, **40**, 726.

Water Pollution Control Federation (1990) *Wastewater Biology: The Microlife*. Special Publication, Alexandria, Va.

Chapter 3
Activated Sludge Floc

MICROBIAL FLOCCULATION IN THE ACTIVATED SLUDGE PROCESS

Microbial flocculation, or *floc* formation, is the aggregation of microbial cells. Effective floc formation is a basic condition for successful treatment using the *activated sludge* process. It is important to have most of the microorganisms present in the biological reactors (*aeration* tanks) converted to flocs, which will then be separated by gravity from the treated *effluent* in the secondary *clarifiers*, leaving behind possibly little residual suspended solids.

Various concepts of floc formation exist. It is believed that floc formation is closely associated with formation of *extracellular polymers* acting through physical *adhesion* and with microbial cells' *electrostatic properties*. Floc formation is most likely a multilevel phenomenon, small clusters of microorganisms with diameters of 6 to 12 μm that perhaps were generated by bacterial growth are strongly connected internally. On the other side, however, the connections among these clusters are porous and not as strong. Microbial *aggregates* may be considered as composite materials. They contain *Zoogleal* and filamentous bacterial cells and extracellular polymers. Physical characteristics of these components are drastically different. Bacterial cells are entities protected by very strong bacterial walls, while extracellular polymers generally do not show noticeable physical integrity. It is possible, however, that various polymers differ substantially in this respect, even within the same aggregate. Specific types of microorganisms are not distributed uniformly in activated sludge flocs but form individual microcolonies (microbial clusters), which sometimes are composed of bacteria that have different *morphologies* than the bacteria in the neighboring microcolonies.

To achieve floc formation, it is important to maintain a proper ratio between the availability of external metabolic substrates and the amount of active microorganisms. At very low substrate loadings, microorganisms use extracellular polymers for their *metabolism* and become dispersed. At low substrate loadings, *filamentous organisms* are favored. They flocculate well, but their settling is slow, and thus clarifier efficiency is adversely affected. At high loadings, microorganisms grow at a constant rate, are very mobile, and do not flocculate.

PHYSICAL CHARACTERISTICS OF FLOCS

The most often studied physical properties of activated sludge flocs include size distribution (*dispersion*), free-settling velocity, porosity, and *permeability*. The size of activated sludge flocs ranges from approximately 1 µm for some dispersed *bacteria* to more than 1500 µm for some large flocs. Most flocs are 100 to 500 µm, depending on *aeration* intensity in the biological reactors. Although flocs smaller than 5 µm are dominant in number, those larger than 50 µm account for most of the surface area, volume, and mass of the activated sludge.

The settling operation in the secondary clarifier can be separated into three zones: free settling in the upper part, flocculating settling in its deeper parts, and compressed settling (*zone settling* and sludge gravity thickening) in the lower part of the clarifier.

During flocculating settling, microbial particles form some aggregates. During zone settling, water is released from the network of growing particle aggregates. As zone settling progresses, the connections between aggregates become more numerous, and larger particle aggregates are formed. Gravity cannot remove all of the water from these large aggregates because there is little difference between the sludge particle and water densities and because of the increasing sludge thickness or *viscosity*. At the zone settling stage, the system no longer consists of particles suspended in water, a *suspension*; a new phase of sludge has been formed.

Porosity of microbial aggregates can be measured by determination of aggregate density, direct microscopic evaluation of thin segments of the aggregates, or use of *confocal microscopy*. In most cases, determination of aggregate density provides only statistical information and does not allow for the determination of the porosity of each sample. The density of bacterial cells typically ranges from 1.02 to 1.10 g/mL, but it is difficult to estimate the exact density or the amount of *polymers* found outside of the cell (extracellular). The *biomass* may have a density only somewhat higher than the density of the liquid phase of the wastewater treated, and its density may not be uniform throughout the total volume. The amount of *extracellular polymers* is typically a small percentage of the total dry mass of microbial aggregates, but it may be as high as 30% of the total volume.

The external surface of activated sludge flocs is more developed (far from being smooth) in the presence of higher levels of metabolic substrates. Moreover, the surface is more developed for larger flocs than for smaller ones, and it is also increased by addition of conditioning polymers, which promote the formation of larger flocs. The external surface development of flocs definitely has an important influence on mass transfer in the activated sludge process, and it also affects settling velocity of flocs and dewatering abilities of the resulting waste sludge.

Table 3.1 shows a summary of physical characteristics of activated sludge flocs.

Composition of Flocs

Microbial flocs are composed of bacterial cells, extracellular polymers, and water that is bound by extracellular polymers and is entrapped in the biomass in a form of pores, channels, and reservoirs. The composition of extracellular polymers is variable and includes, to a significant extent, small threadlike fibers, which together with electrostatic forces are important for microbial flocculation.

Table 3.1 Summary of the physical characteristics of activated sludge flocs.

Characteristics	Values
Floc sizes	
Range, μm	1–1500
Typical, μm	100–500
Sizes of microbial clusters	
Range, μm	6–18
Mean, μm	8
Porosity (as measured by microscopy)	
Range, %	20–59
Mean, %	35
Porosity (as measured by aggregate density)	
Up to, %	99+

In the mixed liquor, some higher organisms are associated with bacterial flocs. They are mostly *protozoa*, *rotifers*, and *nematodes*. The presence or absence of a particular group of these organisms can indicate the expected performance of the process.

Activated sludge flocs are capable of adsorbing some nonbiodegradable components from treated wastewater (such as *heavy metals*) and entrapping some fine suspended solids present originally in the wastewater or introduced to it to improve treatment effects (for example, powdered activated carbon). Only approximately 1 to 10% of the floc mass is viable microbial biomass.

FACTORS PROMOTING FLOC FORMATION

Activated sludge flocs are formed by aeration of *suspensions* of *heterogeneous* microorganisms in the presence of *organic* substrates suitable for their *metabolisms*. To improve floc formation under such circumstances, nutrient requirements should be maintained and *pH* should be kept close to neutral or be slightly *alkaline*.

The most important factor affecting microbial floc formation is the *food-to-microorganism ratio*, that is, the *biochemical oxygen demand* to *mixed liquor volatile suspended solids* loading. However, the nutrient balance, temperature, and *dissolved oxygen* concentration play secondary roles. Water chemistry also influences floc formation in that the number of available bonding sites affects the degree to which organisms adhere to one another.

The use of various *biomass carriers* and *microcarriers*, such as sand particles and other media, in the activated sludge process allows the development of biological fluidized-bed processes, a process whereby liquid or gas moves upward through solid particles at a velocity sufficient to suspend them in fluid; the introduction of various biomass growth support media to biological reactors; and the creation of carrier activated sludge processes. The presence of microcarriers in activated sludge reactors provides a significant surface area (up to 100 000 m^2/m^3 [304 800 sq ft/cu ft] of carriers) for *biofilm* formation and, therefore, creates a high concentration of biomass in the reactors. This high concentration of biomass would not be possible otherwise because the porosity of biofilms is generally much lower than that of activated sludge flocs. In microcarrier reactors, carrier particles coated with biofilm, smaller carrier particles (< 40 μm) entrapped in flocs of relatively large size, some carrier particles free of biofilm, and some flocs free of carriers may be present.

VISUAL OBSERVATIONS OF ACTIVATED SLUDGE AND MICROSCOPIC OBSERVATIONS OF ACTIVATED SLUDGE FLOCS

Over a long period of time, operators who are familiar with the activated sludge from their own plants often develop subjective concepts of how the sludge should look when the process is performing properly. Terms such as "healthy," "fat," or "sick" are used to describe the visual appearance of the sludge. However, because this description is subjective, often each activated sludge process seems unique, and the feel for the appearance of sludge in one plant is not easily transferrable to other plants.

To study morphological characteristics of activated sludge flocs, usually magnifications from 100x to 200x are needed. These observations typically cover shape, dimensions, firmness, and structure of the flocs. The shape of the flocs may be differentiated into somewhat rounded flocs (close to spherical shape) and irregular flocs (usually elongated and built on filaments). The sizes of the flocs can be divided into three basic groups: large flocs of diameters larger than 500 μm, middle-sized flocs of diameters from 150 to 500 μm, and small flocs of diameters less than 150 μm. With the aid of a calibrated *ocular micrometer*, the approximate size of the flocs can be easily estimated. The filamentous organisms protruding from the flocs are not taken under consideration in such estimations. In a properly operated activated sludge process, almost all bacterial cells in the mixed liquor occur as activated sludge flocs. The presence of individual dispersed bacteria often indicates some process upsets. The firmness of the floc is shown by its resistance to *dispersion*. With compact flocs there are only a few open spaces, which are typical for the flocs characterized by an open structure. An open floc structure is often, but not always, caused by the presence of filamentous organisms.

Typical results of microscopic observations of activated sludge *morphology* are given in Figure 3.1. In filling out this form, symbols like "D" (dominant) and "S" (secondary) may be used. An additional form, on morphology of filamentous organisms, may be used for a preliminary identification of these organisms.

CLASSIFICATION OF FLOCS

Typically, activated sludge flocs are classified as ideal, pinpoint, or filamentous. Ideal, non*bulking*, activated sludge flocs are characterized by a balance of a relatively small number of filaments and floc-forming organisms. Relatively strong and large flocs are formed, and a small number of filaments

Size of the flocs:		
	small (< 150 µm)	
	middle (150-500 µm)	
	large (> 500 µm)	
Dispersed cells		
Firmness of the flocs		
	firm	
	weak	
Presence of filaments		
General classifications:		

Figure 3.1 Morphology of activated sludge flocs.

do not interfere with floc settling and compaction. *Supernatant* is clear, and activated sludge shows a relatively low *sludge volume index. Pinpoint (pin) flocs* practically do not contain filamentous organisms. Flocs are relatively small and weak. The supernatant is *turbid* because dense larger flocs settle rapidly and buoyant smaller flocs remain in suspension. Sludge volume index is low. In filamentous bulking of activated sludge, filaments are predominant. Large flocs are formed with extending filaments, which interfere with sludge settling and compaction. A clear supernatant is formed. Sludge volume index is high.

In addition to these main types, activated sludge floc may be classified as dispersed growth, nonfilamentous bulking, rising, and foaming or *scum-forming* flocs. In dispersed growth, organisms do not form flocs and typically do not separate by gravity. Under such conditions, the activated sludge process does not exist any more. In nonfilamentous bulking, which seldom occurs, organisms generate large quantities of extracellular slime. It adversely affects sludge separation and may cause formation of viscous foam. As a consequence of the *denitrification* process, rising sludge occurs in secondary clarifiers from formation of nitrogen gas bubbles. These gas bubbles attach themselves to the flocs and float them to the clarifier surface. Foaming or scum-forming flocs may result from the presence of *Nocardia* spp. and sometimes the presence of *Microthrix parvicella.*

For each specific activated sludge treatment process, individual characteristics of the "proper" activated sludge flocs can be established. These characteristics will be the consequence of the properties of treated wastewater, the modification of the process applied, and existing operating procedures.

SUGGESTED READINGS

Eikelboom, D.H., and van Buijsen (1981) *Microscopic Sludge Investigation Manual.* TNO, Delft, Neth.

Ganczarczyk, J.J. (1983) *Activated Sludge Process. Theory and Practice.* Marcel Dekker, New York.

Jenkins, D., et al. (1986) *Manual on the Causes and Control of Activated Sludge Bulking and Foaming.* Ridgeline Press, Lafayette, Calif.

Li, D.H., and Ganczarczyk, J.J. (1990) Structure of Activated Sludge Flocs, *Biotechnol. Bioeng.*, **35**, 57.

Water Pollution Control Federation (1990) *Wastewater Biology: The Microlife.* Special Publication, Alexandria, Va.

Chapter 4
Fixed-Growth Processes

INTRODUCTION

Treatment that occurs in a fixed-growth reactor (e.g., *trickling filters* and *rotating biological contactors [RBCs]*) is primarily a function of the organisms that inhabit the unit. Those organisms living in *biofilms* or suspended in the wastewater use substances from the wastewater for various *metabolic* reactions, including energy generation, cell maintenance, growth, and reproduction (Bitton, 1994; Gaudy and Gaudy, 1980; and Metcalf & Eddy, 1991). To understand how a fixed-growth reactor operates or why it malfunctions, the role of the organisms in the reactor must be understood. Unfortunately, relatively little is known about the intricacies of the *microbial* processes occurring within fixed-growth processes because these processes have often been considered unimportant. This chapter presents an overview of what is currently known about the microbiology of *aerobic* fixed-growth reactors. It addresses several topics: biofilm formation and regeneration, steady-state populations, roles of the organisms in reactor processes, comparative characteristics relative to suspended culture systems, nuisance organisms, and *indicator species*. The organisms found in fixed-growth reactors may include *bacteria*, *fungi*, *algae*, *protozoa*, and *metazoa* (micro- and macroscopic multicellular organisms).

BIOFILM FORMATION AND REGENERATION

Most of the research that has been conducted on biofilm formation and regeneration has examined either bacterial or protozoan and metazoan biofilm formation.

BACTERIA. Biofilm formation occurs very rapidly in most aerobic wastewater treatment systems. Although a fixed film is often not visible until after a few days, a microscopic layer develops within minutes to hours after startup. The bacteria making up the biofilm are not fecal in origin, but instead come from soil populations, perhaps by infiltration (Harkness, 1966; Prakasam and Dondero, 1967; and Watson, 1945). The first bacteria that contact the media surface may not adhere to it or, if they do, they may attach only temporarily. After 3 to 4 hours, clusters of attached bacteria arise from single cells to form microcolonies. The initial biofilm then becomes rapidly overgrown by these larger secondary colonizing bacteria. Once secondary colonization is complete, biofilm formation proceeds very rapidly (Ware and Loveless, 1958).

Establishment of the initial colonies on the media in fixed-growth reactors results from attachment of *Gram-negative* and flagellated bacteria. Filaments soon become part of the biofilm, perhaps when some of the initial colonizers assume a different *morphology* (i.e., structure or form). The filaments become entangled and, with an extracellular organic binding substance secreted by the cells, form a thick mat (Eighmy et al., 1983). Figure 4.1 shows formation of wastewater biofilm over a 144-hour period (Eighmy et al., 1983).

The process of biofilm formation, which has been studied extensively, can be divided into three phases (Bryers and Characklis, 1982; Characklis, 1981; Corpe, 1980; and Marshall et al., 1971). In the first phase, an *organic* conditioning layer coats the media surface that will be colonized. This happens very rapidly in wastewater treatment systems. The type of material that is colonized is usually not a factor because the organic coating and subsequent biofilm formation occur equally well on glass, concrete, wood, plastics, and metals (Bryers and Characklis, 1982; Characklis, 1981; and Heukelekian and Crosby, 1956). In the second phase, bacteria come in contact with the organically coated surface by random motion in the water, their own spontaneous self-propelled movement, or turbulent forced transport. Bacterial *adsorption* to the media surface is usually reversible so that the cell may

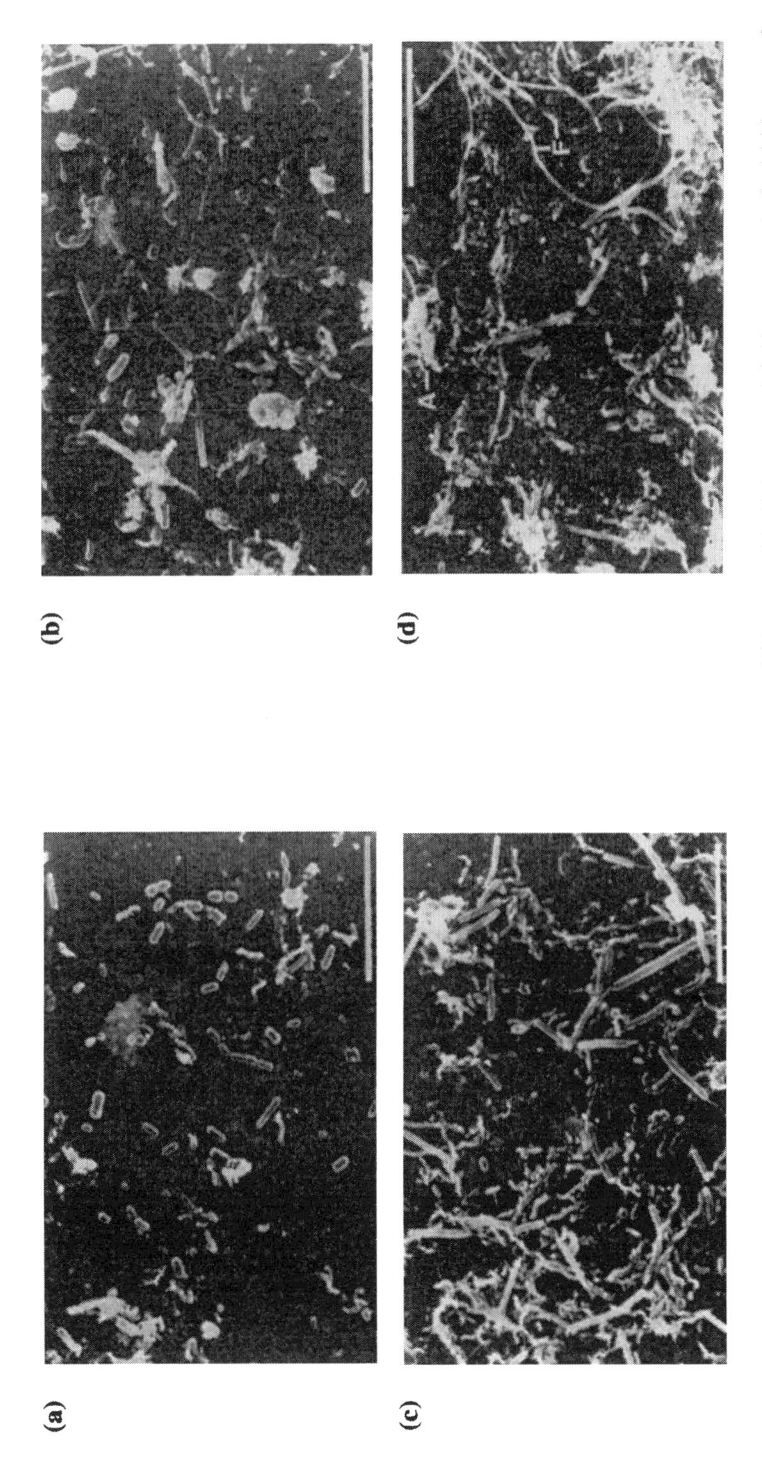

Figure 4.1 Scanning electron micrographs of biofilm development: (a) 24 hours, (b) 48 hours, (c) 96 hours, and (d) 144 hours after incubation in wastewater (bar = 10 μm).

detach. During this phase, adsorption is controlled by weak attractive forces between bacteria and the surface when they are close together and by chemical bonding. The third phase of biofilm formation involves a permanent, irreversible *adhesion* in which the cell secretes a *glycocalyx* (mucouslike *extracellular polymer*) that binds it to the media surface. Individual cells will then grow and reproduce, forming localized microcolonies consisting of similar cells. As growth, reproduction, and more *colonization* occur, the biofilm becomes thicker and visible.

The glycocalyx, which attaches cells to each other and to the media surface, is secreted by the cells (Costerton et al., 1978). It is a sticky structure surrounding the cells and serves many functions, including adhesion, protection, carbon storage, and *ion exchange* (Joyce and Dugan, 1970). It is the glycocalyx that actually forms the zoogleal masses characteristic of wastewater treatment biofilms and *flocs*. Many types of glycocalyxes exist within biofilms (Eighmy et al., 1983), but two major forms occur in wastewater treatment systems. During startup, a branched fingerlike mass consisting mainly of the genus *Zoogloea* predominates (Unz and Dondero, 1967). This branched *Zoogloea* is not very active biochemically. During normal steady-state operation, a more *amorphous* mass that contains a diverse group of bacteria is present and biochemically active.

Biofilms normally develop to a steady-state thickness of 0.5 to 100 mm in fixed-growth reactors, depending on their location within the unit; those exposed to the highest organic loading are thicker. Factors that control the maximum thickness include *hydraulic shear* and limitations on the ability of oxygen and carbon dioxide to reach the inner portion of the biofilm. The length of time a given piece of biofilm remains at steady-state thickness and attached to the media surface has not been quantified accurately, even though it may vary from weeks to months in some units.

Hoehn and Ray (1973) observed that, as the diffusion of oxygen and organic carbon becomes impeded, these essential substrates will not penetrate to the inner portion of the biofilm. Inner regions, therefore, become progressively more *anaerobic*. At some point, the base layer of cells attaching the biofilm to the media dies because of a lack of *substrate*, and *sloughing* occurs. Some researchers have suggested that, once the biofilm reaches a certain thickness, hydraulic shearing causes it to be torn from the media (Antonie, 1976). There remains some controversy about the ultimate cause of sloughing.

Whatever governs the amount and rate of sloughing, it is clear that recolonization of the media surface immediately begins in a process similar to that of biofilm development. The rate at which the biofilm develops on the media surface in a fixed-growth reactor is a function of existing environmental parameters, with temperature and availability of the needed substrates (including oxygen) being the most important.

PROTOZOA AND METAZOA. The *succession* or rate at which these organisms establish themselves in aerobic fixed-growth reactors has been observed and seems to follow a pattern typical of that in *activated sludge*

(Chung, 1987; Chung and Strom, 1991; Kinner, 1983 and 1984; Kinner and Curds, 1987; and Krell, 1980). Within a few days of startup of the system, *flagellates* and small *amoebae* appear throughout the units, followed by *ciliates* such as *Colpidium* and *Glaucoma*. The ciliates are free-swimming protozoa that tend to be predominant through the startup period and then decrease in number as steady-state operation approaches. *Nematodes*; stalked *peritrichs* (e.g., *Opercularia* and *Vorticella*); *rotifers*, large amoebae; and, finally, *carnivorous* ciliates (e.g., *Trachelophyllum* and *Podophrya*) appear. The rate at which the distinct transition to steady state occurs is a function of the organic loading rate to the fixed-growth reactor. At lower loading rates or in the effluent end of the unit, the succession is accelerated because conditions more favorable to protozoa and metazoa exist.

The factors responsible for this succession currently are not known. Researchers disagree on whether environmental parameters such as *dissolved oxygen*, temperature, *pH*, and ammonia; the organisms' *competition* for food; the availability of food; the organisms' efficiencies at processing it; or their reproductive rates are most important.

The major food source for most protozoa and metazoa inhabiting wastewater treatment systems, including fixed-growth reactors, is *enteric* bacteria suspended in the waste stream and not the bacteria composing the biofilm (Curds and Fey, 1969). This is why most protozoa and metazoa live on the surface of the biofilm where they are exposed to their food source and oxygen. Flagellates, the first protozoa to appear in a fixed-growth reactor after startup, usually exist where there are fairly high numbers of bacteria because, to maintain a higher concentration of *prey*, they do not filter as much of the bulk liquid as do the ciliates. The larger, free-swimming *bacterivorous* ciliates (e.g., *Colpidium*) that appear next have relatively simple oral structures. Although they feed more efficiently than the flagellates, they still do not filter as much water as some other ciliates. Thus, they thrive in areas with higher bacterial concentrations.

The stalked protozoa (peritrichs) have oral structures that enable them to filter bacteria out of large volumes of water by generating vortices. There seems to be a trend within the stalked bacterivorous ciliates; those with smaller-diameter oral cavities that filter less water appear where free bacterial concentrations are somewhat low, and those with large oral cavities appear where bacterial concentrations are very low. *Spirotrich* ciliates and rotifers, which are motile, also have highly efficient oral structures, which enable them to survive when bacterial concentrations are very low.

Carnivorous protozoa, including ciliates and amoebae, and metazoa are usually only present in low numbers and only in regions of the treatment units inhabited by their prey. Very few protozoa and metazoa graze on the biofilm itself, probably because most protozoa and metazoa lack the oral structures needed to feed on bacteria in glycocalyx. However, some protozoa (e.g., amoebae) can feed on surface-associated bacteria, and some (e.g., *Chilodonella*) can consume filaments.

STEADY-STATE BIOFILM POPULATIONS

TRICKLING FILTERS. The majority of the microbial research that has been conducted on wastewater biofilms has been performed on trickling filters, probably because they are one of the oldest and most widely used treatment processes. Table 4.1 is a compilation of all of the bacteria, fungi, and algae that have commonly been observed in trickling filters. The surface of a trickling filter is typically composed of a mixture of algae, bacteria, and fungi. Bacteria and fungi predominate near the *influent* end of the filter, while deeper within the reactor near the *effluent* end of the reactor, nitrifying bacteria may thrive. Viewed microscopically, the bacteria on the surface of trickling filter biofilms exist as colonies of rods and *cocci*, which form hollow cylinders and columns with a large surface area (Mack et al., 1975). Deeper in the films, spherical colonies of rods are common.

The *plug flow* nature of trickling filters results in a *stratification* of the protozoa and metazoa within the reactor. Table 4.2 shows the organisms that have commonly been found in trickling filter biofilms. The species of protozoa and metazoa present vary as a function of organic loading rate, depth in the reactor, and season (Barker, 1942 and 1946; Hawkes, 1963; and Wilderer et al., 1982). Unlike many other wastewater treatment processes, trickling filters may have a wide variety of macroscopic metazoa inhabiting the reactor, including *annelid* worms, flatworms, nematodes, rotifers, water

Table 4.1 Bacteria, fungi, and algae commonly found in trickling filters.

Acinetobacter[b]	*Oscillatoria*[a]
Alcaligenes[b]	*Penicillium*[f]
Ascoidea rubescens[f]	*Phormidium*[a]
Bacillus sp.[b]	*Pseudomonas*[b]
Bacillus cereus[b]	*Sepedonium*[f]
Beggiatoa[b]	*Sphaerotilus natans*[b]
Chlamydomonas[a]	*Spirillum*[b]
Chlorella[a]	*Spirochaetes*[b]
Chlorococcum[a]	*Sporotrichium*[f]
Flavobacterium[b]	*Stigeoclonium*[f]
Fusarium aqueductum[f]	*Streptomyces*[f]
Geotrichium[f]	*Subbaromyces splendens*[f]
Monostroma[a]	*Ulothrix*[a]
Nocardia[b]	*Zoogloea ramigera*[b]
Oocystis[a]	

[a] Algae.
[b] Bacteria.
[f] Fungi.

Table 4.2 Protozoa and metazoa commonly found in trickling filters.

Actinophrys[a]	*Litonotus*[c]
Amoeba[a]	*Lumbricus*[w]
Amphileptus[c]	*Metopus*[c]
Arcella[a]	*Monas*[z]
Aspidisca[c]	*Oikomonas*[z]
Bodo[z]	*Opercularia*[c]
Carchesium[c]	*Oxytrichia*[c]
Cercobodo[z]	*Paramecium*[c]
Chilodon[c]	*Peranema*[z]
Chilodonella[c]	*Philodina*[r]
Cinetochilum[c]	*Plagiophyla*[c]
Colpidium[c]	*Pleuromonas*[z]
Colpoda[c]	*Podophrya*[c]
Cyclidium[c]	*Rabdites*[n]
Dero[w]	*Rotaria*[r]
Difflugia[a]	*Spirostomum*[c]
Diplogaster[n]	*Stentor*[c]
Epiphanes[r]	*Tetrahymena*[c]
Epistylis[c]	*Trachelophyllum*[c]
Euglena[z]	*Trepanomonas*[z]
Glaucoma[c]	*Trinema*[a]
Habrotrocha[r]	*Tubifex*[w]
Hartmanella[a]	*Uronema*[c]
Lecane[r]	*Vahlkampfia*[a]
Leucophrys[c]	*Vorticella*[c]

[a] Amoeba.
[c] Ciliate.
[n] Nematode.
[r] Rotifer.
[w] Worm.
[z] Zooflagellate.

mites, *polychaetes*, and snails. Numerous insects and their *larvae* (members of the orders Collembola, Coleoptera, and Diptera) may also be found, and these are often considered nuisance organisms. The distribution of the protozoa and metazoa in trickling filters varies vertically (Wilderer et al., 1982). As the influent organic loading rate varies to the trickling filter, the distribution of ciliated protozoa changes. Most species are present throughout the reactor when the loading rate is low. At higher loading rates, however, some species are present in only limited sections. When considering the interactions among the organisms inhabiting trickling filters, the influent waste is transferred first to the primary decomposers (the bacteria), which are in turn consumed by a variety of *predators*, including the protozoa and metazoa.

Table 4.3 Bacteria and fungi commonly found in RBCs.

Aerobacter aerogenes[b]	*Micrococcus*[b]
Alcaligenes[b]	*Nocardia*[b]
Athrobotrys[f]	*Penicillium*[f]
Bacillus sp.[b]	*Pseudomonas* sp.[b]
Bacillus cereus[b]	*Pseudomonas denitrificans*[b]
Beggiatoa[b]	*Sphaerotilus* sp.[b]
Cladothrix[b]	*Sphaerotilus natans*[b]
Desulfovibrio[b]	*Staphylococcus*[b]
Escherichia coli[b]	*Streptococcus*[b]
Flavobacterium[b]	*Zoogloea filipendula*[b]
Fusarium[f]	*Zoogloea ramigera*[b]
Geotrichium[f]	

[b] Bacteria.
[f] Fungi.

ROTATING BIOLOGICAL CONTACTORS. The bacteria and fungal *biomass* and species distribution in RBCs is similar to that in trickling filters (Table 4.3). However, few algae are observed because the disks are usually covered. Nitrifying bacteria may be found in the later compartments (Sudo et al., 1977). Protozoa and metazoa commonly found in RBCs are shown in Table 4.4.

MICROBIAL ECOLOGY OF AEROBIC FIXED-GROWTH REACTORS

All aerobic fixed-growth reactors have one common feature: their major component is a biofilm (composed of a variety of organisms), which acts as a unit to remove *soluble* and *colloidal* organic compounds from the waste stream. The rate at which the organisms remove these influent compounds is controlled by many physical and chemical events that occur in the gas and liquid phases overlying the organisms. Thus, all of the organic and inorganic substances the cell requires for maintenance and growth must diffuse through the gas and/or liquid phases before contacting the biofilm. In the case of organisms not living on the surface of the film, these substances must also diffuse through the biofilm to reach them. Because the rate of *diffusion* of oxygen to the biofilm is slow in relation to the high demand of the organisms, many of the organisms living in the inner layers do not receive an adequate oxygen supply and thus metabolize anaerobically. Data from various sources indicate that oxygen may only be available in as little as the upper 200 μm of the biofilm (Chen and Bungay, 1981; Hoehn and Ray, 1973; and Swilley, 1965).

Table 4.4 Protozoa and metazoa commonly found in RBCs.

Amoeba[a]	*Metopus*[c]
Amphileptus[c]	*Monas*[z]
Arcella[a]	*Nuclearia*[a]
Aspidisca[c]	*Oikomonas*[z]
Bodo[z]	*Opercularia*[c]
Carchesium[c]	*Oxytrichia*[c]
Cercobodo[z]	*Paramecium*[c]
Chaetogaster[w]	*Peranema*[z]
Chilodonella[c]	*Philodina*[r]
Cinetochilum[c]	*Plagiophyla*[c]
Colpidium[c]	*Pleuromonas*[z]
Colpoda[c]	*Podophrya*[c]
Cyclidium[c]	*Pseudoglaucoma*[c]
Dero[w]	*Proales*[r]
Difflugia[a]	*Rhabdites*[n]
Diplogaster[n]	*Rhabditolaimus*[n]
Epiphanes[r]	*Rotaria*[r]
Epistylis[c]	*Spathidium*[c]
Ethmolaimus[n]	*Spirostomum*[c]
Euglena[z]	*Tachysoma*[c]
Euplotes[c]	*Trachelophyllum*[c]
Glaucoma[c]	*Uronema*[c]
Heterophrys[a]	*Vorticella*[c]
Lecane[r]	*Zoohamnium*[c]
Litonotus[c]	

[a] Amoeba.
[c] Ciliate.
[n] Nematode.
[r] Rotifer.
[w] Worm.
[z] Zooflagellate.

BACTERIA AND FUNGI. The biofilm bacteria, which are responsible for removing the majority of the organics in the reactor, require three types of substances to support their metabolism: an *electron donor*, which can be of any of a variety of organic molecules or inorganic compounds (e.g., ammonia); an *electron acceptor*; and a variety of *macro-* and *micronutrients*. A diverse number of substances may serve as electron acceptors. The choice of acceptor is what determines the type of metabolism the organisms will undergo within the biofilm. Typical electron acceptors are oxygen (*aerobic respiration*), sulfate (*sulfate reduction*), hydrogen or carbon dioxide (*methanogenesis*), other organic molecules (*fermentation*), and nitrate (*denitrification*). The last four processes occur in anaerobic (no free oxygen is available) or *microaerophillic* (very low concentrations of free oxygen) conditions. The required nutrients (e.g., nitrogen and phosphorus) are typically present in excess in municipal wastewater. However, in some industrial wastewater, nutrients may be limited.

Given a choice of all of the electron acceptors listed above, bacteria will use oxygen, nitrate, sulfate, hydrogen or carbon dioxide, or other organics (in that order). This preference for oxygen results from the fact that aerobic respiration yields the most energy of any form of metabolism.

Some of the organic material in the waste stream is colloidal or *particulate* matter. Unlike dissolved organics, these substances must first be adsorbed to the biofilm surface, typically to the glycocalyx, and then degraded to sizes small enough to diffuse through the pore spaces in the biofilm to a cell's surface. These organic molecules must then be solubilized and transported across the cell wall and cell membrane and into the cell where they are metabolized or stored. A similar pathway must be followed by the required nutrients and electron acceptors. Conversely, waste products or end products generated by the cell, including carbon dioxide, carbonate, hydrogen sulfide, and/or ammonia, must diffuse out of the film and into the waste stream. These wastes must be removed from the area adjacent to the cell to prevent the accumulation of *toxic* products (e.g., ammonia), which may inhibit metabolism. It is widely assumed that this *inhibition* does not occur because the waste products, being small molecules, diffuse out of the film very easily or are used within the film by other bacteria. This inhibition effect has not been examined in detail. However, accumulation of toxic products may result in a decrease in internal pH, which could affect treatment efficiency (Harremoës, 1978).

If the amount of carbon, nitrogen, phosphorus, and sulfur required for cell maintenance and growth (i.e., the C:N:P:S ratio) is examined, it is clear that bacteria involved in aerobic degradation of wastewater are limited by the availability of organic carbon, which is consistent with the fact that organic removal (e.g., reduction of biochemical oxygen demand [BOD]) is the primary purpose of secondary treatment (LaRiviere, 1977). During actual operation, aerobic degradation of organic carbon in the wastewater is probably limited by the availability of oxygen within the biofilm.

Little is known about the actual *microbial ecology* of biofilms in aerobic fixed-growth reactors. From visual examination of the biofilm, there is a transition from the influent to the effluent end of the reactors, biofilms on the influent end tending to be thicker and darker brown–black in color. In trickling filters, these films may contain substantial amounts of algae, which appear green, especially where light levels are high. Biofilms in the effluent end of the reactor are thinner and brown–tan in color. If significant amounts of nitrification are occurring, the film may appear orange–brown.

Biofilms in aerobic fixed-growth reactors are composed of zoogleal masses and filaments. Unlike their problematic role in suspended growth treatment systems (e.g., activated sludge processes), filaments may stabilize the fixed-growth reactor biofilms in addition to removing organic matter. In addition, their ability to rapidly colonize surfaces while the *filamentous organisms* are in their flagellated free-swimming unicellular stage may make them integral parts of the biofilm regeneration process (Kinner et al., 1983).

Some data obtained for RBC biofilms suggest that there is a transition in the *physiological* (biological and chemical) and ecological (environmental) conditions to which biofilm bacteria are exposed within an aerobic fixed-growth reactor (Kinner and Maratea, 1984, and Kinner et al., 1985). At high organic loading rates, the bacteria store large amounts of *poly-ß-hydroxybutyrate (PHB)* and *polyphosphate*. The bacteria store PHB when organic carbon is available in excess, while oxygen is limiting in their environment (Senior et al., 1972, and Shively, 1974). These biofilms may also contain bacteria that prefer anaerobic conditions because, when exposed to high organic loadings, there may be substantial anaerobic portions within the film. As loading rates decrease (based on samples taken toward the effluent end of the reactor), the amount of PHB and polyphosphate stored decreases. In compartments with very low loading rates, *prosthecate bacteria*, which have numerous appendages, may predominate. These bacteria are known to use a wide variety of organics (Staley, 1971, and Stanley et al., 1979), which may explain their presence in the latter portion of a reactor where *refractory organics* are likely to be the primary organic carbon sources remaining.

If *nitrification* is occurring in the reactor, usually after organic removal is fairly complete, the biofilm may contain predominately nitrifying bacteria, which possess a large amount of internal membrane (*cytomembrane*) (Kinner et al., 1985). These membranes are the sites within the bacteria where ammonia is oxidized to nitrite and nitrite is oxidized to nitrate to generate the energy used to transform inorganic carbon to organic carbon.

There is probably a "division of labor" among the biofilm bacteria affecting wastewater treatment in an aerobic fixed-growth reactor (LaRiviere, 1977). The bacteria normally found in these biofilms are *facultative anaerobes* (i.e., capable of changing their metabolism from an aerobic to anaerobic mode). The factor that controls their metabolism is believed to be the energy yield of the process.

Within the biofilm of an aerobic fixed-growth reactor, processes occur as follows. At the influent end of the reactor, or where organic loading rates are high, oxygen is rapidly used to degrade the large amount of organic matter present. If nitrification does occur in the upper layers of the biofilm, the nitrate generated will be used for denitrification—the most energy-yielding anaerobic mode of respiration. Strand et al. (1985) found organisms capable of denitrification composing 7% of fixed growths operating under certain conditions. If the wastewater contains hydrogen sulfide (H_2S) or if H_2S is generated within the anaerobic layers of the film from sulfate reduction, there may be sulfur storage in some of the cells in the upper aerobic layers. The inner anaerobic layers of the film are likely to have bacteria that perform either sulfate reduction and/or fermentation to derive energy from organic degradation. Small amounts of methanogenic bacteria may exist in the innermost layers of the biofilm. This is true even though the biofilm is growing in an "aerobic" reactor.

The organic loading decreases as the wastewater flows through it and, as a result, the pattern and depth of the zonation within the biofilm changes. Although the same amount of oxygen may be available, the organic carbon concentration has decreased, resulting in a larger percentage of bacteria performing aerobic respiration. Conversely, methanogenesis, sulfate reduction, and fermentation will decrease until these processes are virtually eliminated by the effluent end of the unit. The prosthecate bacteria, which can use refractory organics, and the nitrifying bacteria thrive in these effluent-end biofilms.

PROTOZOA AND METAZOA. Although there is still controversy surrounding the major role of protozoa and metazoa in aerobic treatment processes, most of the evidence indicates that these organisms remove the dispersed bacteria, especially the *enteric* species, from the wastewater (Curds and Fey, 1969; Curds et al., 1968; and Gude, 1979). By doing this, they decrease the effluent BOD and suspended solids concentrations. They may also play several other roles in treatment. Bacterial *mineralization* of organic carbon (conversion to inorganic form) and respiration may be stimulated by protozoa grazing on bacteria (Hunt et al., 1977). Protozoa and metazoa may also be responsible for the regeneration of nutrients within a reactor (Johannes, 1965; Nilsson, 1977; and Stout, 1973).

There is a transition in the protozoan and metazoan community within a fixed-growth aerobic process similar to that observed for bacteria (Kinner, 1983, and Krell, 1980). The free-swimming protozoa and metazoa are typically found near the influent end of the reactor. Those species that can filter more water and have more complex oral structures, many of which are attached to the biofilm surface, flourish near the effluent end. Carnivorous protozoa and metazoa are present in biofilms where their prey exist in substantial numbers. This distribution, which was discussed in the earlier "Biofilm Formation and Regeneration" section, may be explained in a manner analogous to protozoan and metazoan succession.

COMPARISON WITH SUSPENDED GROWTH TREATMENT PROCESSES. A comparison of the bacterial *species* found in suspended growth treatment and fixed-growth processes will yield few differences. The suspended growth flocs tend to have fewer, if any, of the more strictly anaerobic microorganisms. This is not surprising because there is much less of a tendency to develop long-term anaerobic conditions within the flocs under typical operating conditions.

The problem of mass transfer of substrates to the cells may be much less important in flocs because diffusion is multidirectional, whereas in fixed growths the movement is more unidirectional. The mixing effect induced by aeration also tends to reduce the size of the stagnant boundary layer in suspended growth processes, which decreases mass-transfer resistance. These conditions may result in less microbial stratification within the flocs.

One of the major differences between the two treatment processes is the role of the filaments. Most research indicates that large amounts of filaments, such as *Sphaerotilus*, are nuisance organisms in suspended growth flocs because they can hinder secondary settling, often resulting in the condition known as "*bulking*." However, in fixed-growth processes, *Sphaerotilus* has been identified as a predominant constituent of healthy biofilms (Kinner et al., 1983). In these films, *Sphaerotilus* may not only remove substantial amounts of organic carbon but may also serve as a structural member maintaining the integrity of the film. In addition, *Sphaerotilus*' ability to form a flagellated single-cell organism may aid in the rapid recolonization of the media following sloughing (the process when a piece of fixed growth falls off the media). It seems that the typically excellent settling abilities of the dense sloughed fixed growth during secondary clarification are not affected by the presence of the filaments.

The protozoa and metazoa that develop in suspended growth flocs tend to be less specialized than those observed in fixed-growth processes. This lack of specialization has made the use of indicator schemes for suspended culture treatment less attractive. The protozoa and metazoa inhabiting the mixed liquor flocs are exposed to a wide range of environmental parameters because these processes usually are operated with recycled sludge. As a result, the organisms tend to be less specialized and more tolerant than those found in traditional *plug-flow*, fixed-growth reactors in which sludge recycle is less typical and biofilm organisms are confined to a certain area of the reactor.

NUISANCE ORGANISMS

The ecology of many of the nuisance organisms within fixed-growth processes is not well known. Generally, these organisms are present during periods of unsatisfactory operation, even though, in some cases, they may not have any effect on organic removal. Once established, they are usually very difficult to eliminate.

BACTERIA AND FUNGI. Probably the least well understood nuisance organisms are the bacteria and fungi, which may invade fixed growths and exclude other species associated with normal treatment. Although fungi may be a constituent of trickling filter biofilms during normal operation (Cooke and Hirsch, 1958, and Tomlinson and Williams, 1975), excessive fungal growth causes ponding on the filter's surface. The fungal-dominated films tend to be associated with strong wastewater and high organic loading (Hawkes, 1963) and often a lower pH (5.5 to 6). Gray (1983), however, found *Subbaromyces splendens* growing in a trickling filter treating weak domestic wastewater. Generally, fungal growth is absent in summer and predominates in winter or spring (Haenseler et al., 1923). The excessive weight of the thick fungal films may cause failure of plastic media filters (Gray, 1983) and RBC shafts.

FILAMENTS. In some cases, a shaggy white filamentous biofilm develops with an underlying black layer (Chesner and Iannone, 1980). The white filaments are usually labeled as *Beggiatoa*, although the true genus present is often not determined. These bacteria seem to thrive where an underlying zone of sulfate-reducing bacteria has developed. Alleman et al. (1982) have associated *Desulfovibrio*, a major sulfate-reducing bacterium, with the black metal-sulfide precipitating layer of organically overloaded biofilms. In these conditions, the sulfate reducers, which are anaerobic, thrive because oxygen is rapidly depleted in the upper layers of the biofilm. Organic carbon diffuses through the film to the organisms, as does sulfate, which is normally present in substantial amounts in domestic wastewater. These bacteria may generate H_2S as a product of sulfate reduction and the degradation of sulfur-containing organics. The H_2S released by these organisms will diffuse up through the film and may be used by filamentous organisms, such as *Beggiatoa*, which use oxygen and transform H_2S to elemental sulfur. These filaments usually live at the interface between the aerobic and anaerobic zones.

Once this type of biofilm *community* develops, a series of undesirable operating conditions may occur, including odors, decreased organic removal efficiency, and development of excessively heavy growth. In RBCs, this last condition is particularly detrimental because it may result in excessive weight and, if unchecked, could lead to shaft failure. Growth of this type of biofilm can be very difficult to prevent unless the cause of the problem is eliminated and the existing cells are removed by dosing with a strong oxidant or scouring with water. Another technique found to be effective for those systems in which waste concentrations of sulfide are high is to reduce sulfide below 5 mg/L and at the same time reduce the organic loading below 0.01 kg/m^2 (2 lb BOD/1000 sq ft). Some possible causes of the problem are organic overloading, low dissolved oxygen concentrations, or high inputs of sulfate or H_2S to the influent. However, there are cases when this type of biofilm has developed where none of these symptoms has apparently existed.

FLIES. Many metazoa, particularly insects, worms, and snails, may become nuisance organisms in fixed-growth reactors. The most predominant of these is the *Psychoda* fly, a mothlike two-winged fly that thrives in the alternating wet/dry conditions found in trickling filters. *Psychoda* larvae may play a significant role in proper filter operation, but the swarming adults present a problem (Painter, 1980). The flies, although not directly detrimental to filter operation, are an annoyance to operators and adjacent residents. There are also some other insect genera such as *Anisopus* and *Achorutes* (the springtail) that may inhabit trickling filters and become pests. Usually, by altering trickling filter operations (e.g., by increasing the amount of recirculation), these problems can be alleviated (Culp and Heim, 1978, and WEF, 1996).

WORMS. *Segmented worms*, including *Lumbricus* and the bristleworm *Dero*, may also be found in fixed-growth reactor biofilms. The worms may clog distribution arms or cause ponding in the trickling filter (Scafidi, 1984). More often, the worms may graze on the biofilm so severely that they significantly deplete the bacterial populations. The causes of these infestations are usually unknown, and eliminating the organisms is difficult (Solbe, 1975).

SNAILS. In some cases, snails have been found in trickling filters, particularly where there is potential for nitrification. These organisms may not create a serious problem if their numbers are low and they graze on the biofilm, especially on the algae on the surface. However, they often pose serious difficulties for filter operation. The main problem associated with them results from their shells. Once the organisms die, the remaining shells clog the trickling filter, block pipes, or abrade pumps (Hawkes, 1963, and Learner, 1975).

TROUBLESHOOTING. Overall, much research still needs to be done, especially at the microbial level, to understand the causes of nuisance growth in fixed-growth reactors. When nuisance organisms proliferate, they tend to dominate the entire biofilm, causing a decrease in organism diversity. The resulting problem is usually very difficult to eradicate and often may require drastic solutions such as complete overhaul of the unit. Most troubleshooting techniques currently available (Culp and Heim, 1978, and WEF, 1996) recommend changes in hydraulic or organic loading or recirculation. In some cases, however, the problem arises in reactors in which accepted loading rates are being used.

USE OF BIOFILM ORGANISMS AS INDICATORS OF REACTOR PERFORMANCE

Protozoa and metazoa have been used in various schemes as indicators of effluent quality in fixed-growth reactors (Curds and Cockburn, 1970, and Kinner, 1983). These organisms are relatively easy to observe using light microscopy, do not usually require extensive laboratory procedures for identification, and are fairly susceptible to changes in environmental conditions. The indicator schemes that have been developed use rapid detection techniques, which are easily learned and use low-cost equipment. Generally, the major focus of these schemes is to ensure a good correlation between the presence of certain protozoa and metazoa and the effluent quality. Typically, the indicator schemes require the operator to examine a specified volume of biofilm or reactor effluent using a light microscope. The operator usually scans the sample and looks for specific protozoa or metazoa and is often required to count the numbers of each type of organism observed. Using this information, the operator can then determine the effluent quality by comparison to the numbers and types of organisms associated with each specific condition (e.g., good, fair, and poor effluent quality or approximate BOD level). Usually the more stalked protozoa (peritrichs) and the greater the diversity of organisms, the better the effluent quality.

Trickling filter indicator schemes have been hampered somewhat by the lack of access to the biofilms with the best indicator potential (i.e., those in the effluent end of the reactor). It is there that the most sensitive protozoan and metazoan species grow. Usually, indicator schemes for trickling filters have involved examination of sloughed biofilm contained in the effluent, which may have originated from various levels in the unit. Despite these problems, Curds and Cockburn (1970) developed a strategy to relate species diversity to effluent BOD concentrations. Wilderer et al. (1982) also found that certain protozoa were present as a function of a trickling filter's organic loading. Several monitoring procedures have been proposed for RBCs (Hoag et al., 1980; Kinner, 1983; Kinner et al., 1988; Krell, 1980; and Sudo et al., 1977). The various schemes require further research for applicability in fixed-growth reactors.

References

Alleman, J.E.; Veil, J.A.; and Canady, J.T. (1982) Scanning Electron Microscope Evaluation of Rotating Biological Contactor Biofilm. *Water Res.* (G.B.), **16**, 543.

Antonie, R.L. (1976) *Fixed Biological Surfaces—Wastewater Treatment.* CRC Press, Cleveland, Ohio.

Barker, A.N. (1942) The Seasonal Incidence, Occurrence and Distribution of Protozoa in the Bacterial Bed Process of Sewage Disposal. *Annu. Appl. Biol.*, **29**, 23.

Barker, A.N. (1946) The Ecology and Function of Protozoa in Sewage Purification. *Annu. Appl. Biol.*, **33**, 314.

Bitton, G. (1994) *Wastewater Microbiology.* Wiley–Liss, New York.

Bryers, J.D., and Characklis, W.G. (1982) Processes Governing Primary Biofilm Formation. *Biotechnol. Bioeng.*, **24**, 2451.

Characklis, W.G. (1981) Bioengineering Report: Fouling Biofilm Development—A Process Analysis. *Biotechnol. Bioeng.*, **23**, 1923.

Chen, Y.S., and Bungay, H.R. (1981) Microelectrode Studies of Oxygen Transfer in Trickling Filter Slimes. *Biotechnol. Bioeng.*, **23**, 781.

Chesner, W.H., and Iannone, J.J. (1980) Current Status of Municipal Wastewater Treatment with RBC Technology in the U.S. *Proc. 1st Natl. Symp. Rotating Biol. Contactor Technol.*, **1**, 53.

Chung, J.C. (1987) Ecology of Filamentous Bacteria and Other Dominant Microorganisms in the Rotating Biological Contactor. Ph.D. dissertation, Rutgers Univ., New Brunswick, N.J.

Chung, J.C., and Strom, P.F. (1991) Microbiological Study of Ten New Jersey Rotating Biological Contactor Wastewater Treatment Plants. *J. Water Pollut. Control Fed.*, **63**, 35.

Cooke, W.B., and Hirsch, A. (1958) Continuous Sampling of Trickling Filter Populations. II. Populations. *Sew. Ind. Wastes*, **30**, 138.

Corpe, W.A. (1980) Microbial Surface Components Involved in Adsorption of Microorganisms onto Surfaces. In *Adsorption of Microorganisms onto Surfaces.* G. Bitton and K.C. Marshall (Eds.), Wiley & Sons, New York, 105.

Costerton, J.W.; Geesey, G.G.; and Cheng, K.J. (1978) How Bacteria Stick. *Sci. Am.*, **238**, 86.

Culp, G.L., and Heim, N.F. (1978) Field Manual for Performance Evaluation and Troubleshooting at Municipal Wastewater Treatment Facilities. EPA-430/19-78-001.

Curds, C.R., and Cockburn, A. (1970) Protozoa in Biological Sewage Treatment Processes. I. A Survey of the Protozoan Fauna of British Percolating Filters and Activated Sludge Plants. *Water Res.* (G.B.), **4**, 225.

Curds, C.R., and Fey, G.J. (1969) The Effect of Ciliated Protozoa on the Fate of *Escherichia coli* in the Activated Sludge Process. *Water Res.* (G.B.), **3**, 853.

Curds, C.R.; Cockburn, A.; and Vandyke, J.M. (1968) An Experimental Study of the Role of the Ciliated Protozoa in the Activated Sludge Process. *Water Pollut. Control*, **67**, 312.

Eighmy, T.T.; Maratea, D.; and Bishop, P.L. (1983) Electron Microscopic Examination of Wastewater Biofilm Formation and Structural Components. *Appl. Environ. Microbiol.*, **45**, 1921.

Gaudy, A.F., Jr., and Gaudy, E.T. (1980) *Microbiology for Environmental Scientists and Engineers.* McGraw–Hill, New York.

Gray, N.F. (1983) Ponding of a Random Plastic Percolating Filter Medium Due to the Fungus *Subbaromyces splendens* Hesseltine in the Treatment of Sewage. *Water Res.* (G.B.), **17**, 1295.

Gude, H. (1979) Grazing by Protozoa as a Selection Factor for Activated Sludge Bacteria. *Microbiol. Ecol.*, **5**, 225.

Haenseler, C.M.; Moore, W.D.; and Gaines, J.G. (1923) Fungi and Algae of the Sprinkling Filter Bed with Special Reference to Their Seasonal Distribution. *Bull. N.J. Agric. Exp. Stn.*, **390**, 39.

Harkness, N. (1966) Bacteria in Sewage Treatment Processes. *J. Proc. Inst. Sew. Purification*, **6**, 542.

Harremoës, P. (1978) Biofilm Kinetics. In *Water Pollution Microbiology.* R. Mitchell (Ed.), Wiley & Sons, New York, **2**, 71.

Hawkes, H.A. (1963) *The Ecology of Wastewater Treatment.* Macmillan, New York.

Heukelekian, H., and Crosby, E.S. (1956) Slime Formation in Polluted Waters. II. Factors Affecting Slime Growth. *Sew. Ind. Wastes*, **28**, 78.

Hoag, G.; Widmer, W.; and Hovey, W. (1980) Microfauna and RBC Performance: Laboratory and Full-Scale Systems. *Proc. 1st Natl. Symp. Rotating Biol. Contactor Technol.*, **1**, 167.

Hoehn, P., and Ray, A. (1973) Effects of Thickness on Bacterial Film. *J. Water Pollut. Control Fed.*, **45**, 2302.

Hunt, H.W.; Cole, C.V.; Klein, D.A.; and Coleman, D.C. (1977) A Simulation Model for the Effect of Predation on Bacteria in Continuous Culture. *Microbiol. Ecol.*, **3**, 259.

Johannes, R.E. (1965) Influence of Marine Protozoa on Nutrient Regeneration. *Limnol. Oceanogr.*, **10**, 434.

Joyce, G.H., and Dugan, P.R. (1970) The Role of Floc-Forming Bacteria in BOD Removal from Wastewater. *Dev. Ind. Microbiol.*, **11**, 377.

Kinner, N.E. (1983) A Study of the Microorganisms Inhabiting Rotating Biological Contactor Biofilms During Various Operating Conditions. Ph.D. dissertation, Univ. of N.H., Durham.

Kinner, N.E. (1984) An Evaluation of the Feasibility of Using Protozoa and Metazoa as an Indicator of RBC Effluent Quality. *Proc. 2nd Int. Conf. Fixed-Film Biol. Process.*, **1**, 74.

Kinner, N.E., and Curds, C.R. (1987) Development of Protozoan and Metazoan Communities in Rotating Biological Contactor Biofilms. *Water Res.* (G.B.), **21**, 481.

Kinner, N.E., and Maratea, D. (1984) Evaluation of RBC Biofilm Bacteria: A Biogeochemical Approach. *Proc. 2nd Int. Conf. Fixed-Film Biol. Process.*, **1**, 21.

Kinner, N.E.; Balkwill, D.L.; and Bishop, P.L. (1983) Light and Electron Microscopic Studies of Microorganisms Growing in Rotating Biological Contactor Biofilms. *Appl. Environ. Microbiol.*, **45**, 1659.

Kinner, N.E.; Maratea, D.; and Bishop, P.L. (1985) An Electron Microscopic Evaluation of Bacteria Inhabiting Rotating Biological Contactor Biofilms During Various Loading Conditions. *Environ. Technol. Lett.*, **6**, 455.

Kinner, N.E.; Curds, C.R.; and Meeker, L.D. (1988) Protozoa and Metazoa as Indicators of Effluent Quality in Rotating Biological Contactors. *Water Sci. Technol.* (G.B.), **20**, 199.

Krell, H.W. (1980) Untersuchgen zur Okologie einer Scheibentropfkor Perkaskude als Modell einer Selbstreinigungsstrecke. Master thesis, Univ. (TH) Karlsruhe, Ger.

LaRiviere, J.W.M. (1977) Microbial Ecology of Liquid Waste Treatment. *Adv. Microbiol. Ecol.*, **1**, 215.

Learner, M.A. (1975) Insecta. In *Ecological Aspects of Used-Water Treatment. Vol. I. The Organisms and Their Ecology.* C.R. Curds and H.A. Hawkes (Eds.), Academic Press, New York, 337.

Mack, W.N.; Mack, J.P.; and Ackerson, A.O. (1975) Microbial Film Development in a Trickling Filter. *Microbiol. Ecol.*, **2**, 215.

Marshall, K.C.; Stout, R.; and Mitchell, R. (1971) Mechanisms of the Initial Events in the Sorption of a Marine Bacteria to Surfaces. *J. Gen. Microbiol.*, **68**, 337.

Metcalf & Eddy, Inc. (1991) *Wastewater Engineering: Treatment, Disposal and Reuse.* 3rd Ed., McGraw–Hill, New York.

Nilsson, R. (1977) On Food Vacuoles in *Tetrahymena pyriformis* G.L. **24**, 502.

Painter, H.A. (1980) *A Survey of Filter Fly Nuisances and Their Remedies.* Tech. Rep. 155, Water Research Centre, Stevenage, Eng.

Prakasam, T.B.S., and Dondero, N.C. (1967) Aerobic Heterotrophic Bacterial Populations of Sewage and Activated Sludge Bacteria. II. Method of Characterization of Activated Sludge Bacteria. *Appl. Microbiol.*, **15**, 1122.

Scafidi, J. (1984) Operational Problems at Salem, NH. *J. New Eng. Water Pollut. Control. Assoc.*, **18**, 72.

Senior, P.J.; Beech, G.A.; Ritchie, G.A.F.; and Dawes, E.A. (1972) The Role of Oxygen Limitation in the Formation of Poly-ß-Hydroxybutyrate During Batch and Continuous Culture of *Azotobacter beijerinckii. Biochem. J.*, **128**, 1193.

Shively, J.M. (1974) Inclusion Bodies of Prokaryotes. *Annu. Rev. Microbiol.*, **28**, 167.

Solbe, J.F.D.L. (1975) Annelida. In *Ecological Aspects of Used-Water Treatment. Vol. I. The Organisms and Their Ecology.* C.R. Curds and H.A. Hawkes (Eds.), Academic Press, New York, 305.

Staley, J.T. (1971) Incidence of Prosthecate Bacteria in a Polluted Stream. *Appl. Microbiol.*, **22**, 496.

Stanley, P.M.; Ordal, E.J.; and Staley, J.T. (1979) High Numbers of Prosthecate Bacteria in Pulp Mill Waste Aeration Lagoons. *Appl. Environ. Microbiol.*, **37**, 1007.

Stout, J.D. (1973) The Relationship Between Protozoan Populations and Biological Activity in Soils. *Am. Zool.*, **13**, 193.

Strand, S.E.; McDonnell, A.J.; and Unz, R.F. (1985) Concurrent Denitrification and Oxygen Uptake in Microbial Films. *Water Res.* (G.B.), **19**, 335.

Sudo, R.; Okada, M.; and Mori, T. (1977) Rotating Biological Contactor Microbe Control in RBC. *J. Water Wastes*, **19**, 855.

Swilley, E.L. (1965) Transport Phenomena and Rate Control in Trickling Flow Models. Ph.D. dissertation, Rice Univ., Tex.

Tomlinson, T.G., and Williams, I.L. (1975) Fungi. In *Ecological Aspects of Used-Water Treatment. Vol. I. The Organisms and Their Ecology.* C.R. Curds and H.A. Hawkes (Eds.), Academic Press, New York, 93.

Unz, R.F., and Dondero, N.C. (1967) The Predominant Bacteria in Natural Zoogloeal Colonies. I. Isolation and Identification and II. Physiology and Nutrition. *Can. J. Microbiol.*, **13**, 1671.

Ware, G.C., and Loveless, J.E. (1958) The Construction of Biological Film in a Percolating Sewage Filter. *J. Appl. Bacteriol.*, **21**, 308.

Water Environment Federation (1996) *Operation of Municipal Wastewater Treatment Plants.* 5th Ed., Manual of Practice No. 11, Alexandria, Va.

Watson, J.M. (1945) The Bionomics of Coprophilic Protozoa. *Biol. Rev.*, **21**, 121.

Wilderer, P.A.; Hartmann, L.; and Nahrang, T. (1982) The Importance of Ecological Considerations on Design and Operation of Trickling Filters. *Proc. 1st Int. Conf. Fixed-Film Biol. Process.*, **3**, 1599.

Chapter 5
Wastewater Stabilization Lagoons

INTRODUCTION

Shallow lagoons—often called wastewater lagoons, *stabilization* lagoons, *oxidation* lagoons, or simply lagoons—use microbial *metabolism* to stabilize the nutrients in wastewater. The stabilization process depends mainly on effective use of *bacteria* to degrade *organic* materials and efficient use of *algae* to maintain an oxygen level in the system adequate for the oxidation reactions required. The process is similar to the natural purification phenomenon that occurs in a stream, natural lagoon, or lake and may include *sedimentation*, *aerobic* and *anaerobic degradation* by *bacteria* in *suspension*, and *algae photosynthesis*.

Dissolved nutrients such as nitrogen and phosphorus are used by algae to produce cellular growth materials. Algae consume carbon dioxide during photosynthesis and release oxygen as an end product. Although operation of a lagoon is biologically complex and depends on many environmental factors (chemical and physical), lagoons have become increasingly more popular for wastewater treatment. Stabilization lagoons or lagoons that are properly designed, constructed, and operated can produce *effluents* that will meet U.S. secondary treatment standards. However, their inability to consistently meet effluent total suspended solids (TSS) limitations is a major problem. Currently, the U.S. Environmental Protection Agency and many states are beginning to strenuously address this problem.

Lagoons can achieve high treatment efficiencies with low operational costs for a wide variety of wastewaters. However, limitations on their use include the need for land and the potential for nuisance odors. Poor performance may result from overloading and ice cover in cold climates.

TYPES OF STABILIZATION LAGOONS

Wastewater lagoons are designed as a complete self-contained treatment process. They are constructed facilities of earthen materials. Size and depth are the primary design parameters, and they may be designed to treat wastes from single households or large cities. There are many different applications; the following are the common types of lagoons in use.

CONVENTIONAL LAGOONS. Conventional lagoons, often called stabilization lagoons, receive raw wastewater without prior treatment. A conventional lagoon is a secondary treatment facility designed to reduce solids and *biochemical oxygen demand (BOD)* through settling and bacterial decomposition. The required *dissolved oxygen (DO)* of the system is primarily supplied by photosynthetic algae living in the lagoon.

OXIDATION/POLISHING LAGOONS. The oxidation lagoon primarily provides additional treatment after a primary treatment process. These lagoons are commonly used in series. Polishing lagoons are also used to further reduce BOD and suspended solids (SS) in secondary treated effluent before discharge. They are considered to be tertiary treatment.

STORAGE LAGOONS OR CELLS. Storage lagoons or cells are used to contain industrial wastes before a treatment process or land application. In municipal wastewater treatment processes, storage cells in series are often used to improve effluent quality. They also provide flexibility to meet discharge requirements.

MECHANICALLY ASSISTED LAGOONS. Mechanically assisted lagoons are designed with recirculation capability. Circulation provides mixing of lagoon contents. This design is used for high-strength industrial waste as well as nutrient removal from municipal wastewater.

MICROBIAL CLASSIFICATION OF STABILIZATION LAGOONS

Lagoon systems or lagoons may also be classified according to the oxygen relationships of the organisms that are metabolizing the organic waste. Aerobic organisms require molecular oxygen (O_2) for respiration or *bio-oxidation* (the process by which organisms and all living things obtain energy for *metabolic* processes). These organisms use DO in the water. Anaerobic and *anoxic* organisms use compounds other than oxygen as *electron acceptors* in energy-generating reactions. Nitrate (NO_3^-) or sugar compounds ($C_6H_{12}O_6$) are commonly reduced in anaerobic and fermentative processes.

Facultative anaerobes, which are the most abundant type of organisms in wastewater, can use either dissolved or combined oxygen for their metabolic activities. This versatility occurs because of the diverse respiratory pathways that they can use. There are three types of stabilization lagoon systems determined by oxygen characterization.

AEROBIC LAGOONS. Aerobic lagoons have oxygen distributed throughout the depth of the lagoon. The necessary oxygen usually must be supplied by mechanical agitation or air diffusers. This category includes, by design, complete-mix aerated lagoons, shallow oxidation lagoons, and polishing/storage lagoons (Figure 5.1).

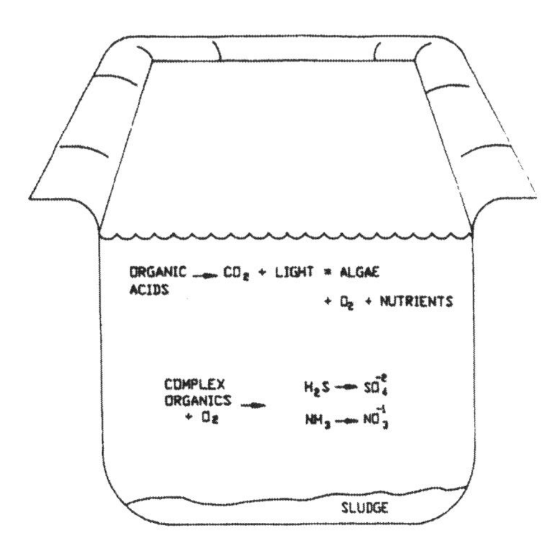

Figure 5.1 Aerobic lagoon.

ANAEROBIC LAGOONS. Anaerobic lagoons are devoid of oxygen and treatment depends on a process of anaerobic metabolism, which generates carbon dioxide, methane, and hydrogen. These lagoons are used primarily for industrial wastewater. Badly overloaded municipal lagoons may become anaerobic with the production of hydrogen sulfide and other carbonaceous gases (Figure 5.2).

FACULTATIVE LAGOONS. Facultative lagoons are the most common type in use today. These systems have three zones: an aerobic layer at the top, a facultative transition zone, and an anaerobic layer at the bottom. The algae are responsible for supplying most of the oxygen. Facultative bacteria that are present continue decomposition activities when oxygen conditions change (Figure 5.3).

BIOLOGICAL ACTIVITY IN WASTEWATER STABILIZATION

The physical and biochemical processes occurring simultaneously in a lagoon are complex; however, they may be summed up in terms of an energy balance. Energy (food) for the naturally occurring organisms in wastewater enters the lagoon with the *organic matter* present in the wastewater. Sunlight also enters, and its energy is trapped by *chlorophyll* molecules present in the algae.

Settleable solids from *influent* wastewater fall to the bottom of the lagoon and are slowly decomposed by anaerobic bacteria. This process, called *fermentation*, generates gases such as methane, nitrogen, hydrogen sulfide,

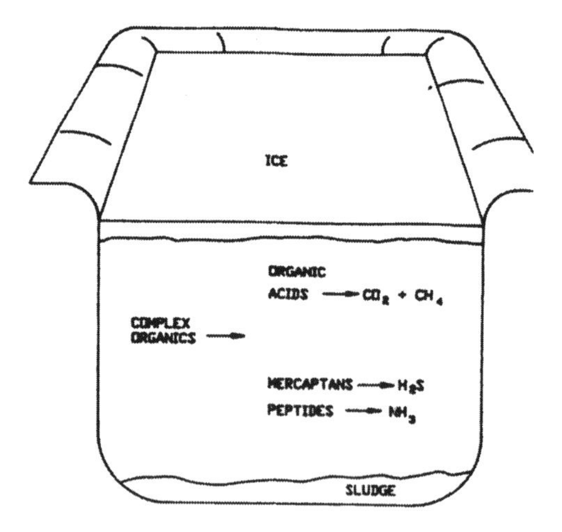

Figure 5.2 Anaerobic lagoon.

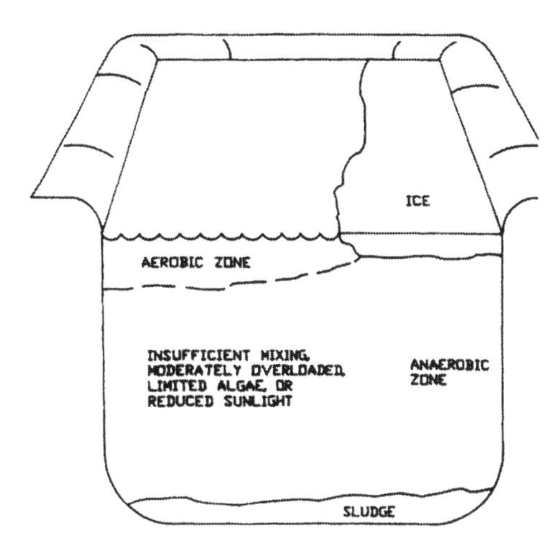

Figure 5.3 Facultative lagoon.

and carbon dioxide, which are released to the upper layers. Also, the dissolved chemicals of the decomposed solids are released to the upper zone of the lagoon, and there is a buildup of bottom sludge. The degradation of bottom sludge is temperature dependent and slows greatly during cold winter months. Although *biodegradation* of organics may be carried out aerobically or anaerobically (Figure 5.4), the result is the same—the release of energy is necessary to drive the life processes for all living cells. In the process, organics are removed from wastewater and converted to substances that are no longer biodegradable. Because *aerobic respiration* is more efficient than *anaerobic respiration* in a lagoon, the algae, which produce molecular oxygen as a product of photosynthesis, are very important to the system.

The process of stabilization is a mutually beneficial interaction (or *symbiosis*) among bacteria and algae. The aerobic bacteria need oxygen to degrade organics for the release of energy while the algae release oxygen during photosynthesis. The algae use the carbon dioxide produced during respiration by the bacteria to carry on photosynthesis.

The proper functioning of waste stabilization lagoons involves the interaction of solar energy, algae, and bacteria (Figure 5.4). *Aerobic oxidation* converts organics such as carbohydrates to bacterial cells, carbon dioxide, water, and energy. In the presence of light and nutrients (phosphorus and nitrogen), algae convert carbon dioxide to cells and free oxygen through the photosynthesis mechanism. This is essentially the reverse of the preceding reaction. These two simplified bacterial and algae transformations of carbohydrates may occur. In this case, the end products are organic acids and carbon dioxide. Sometimes, these organic acids are further decomposed to methane with the growth of methane bacteria. The interactions of environmental factors and major biological reactions are shown in Figure 5.4.

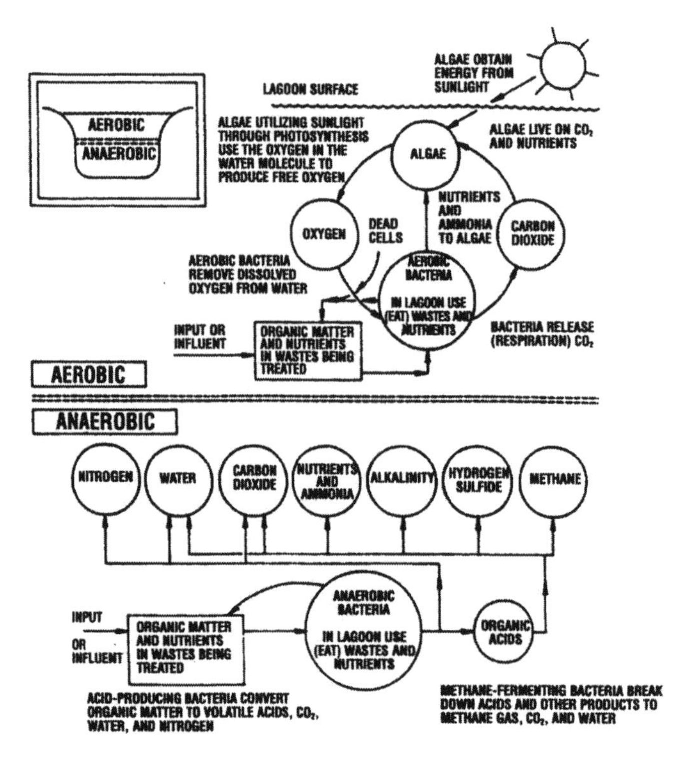

Figure 5.4 Process of decomposition in aerobic and anaerobic layers of a lagoon.

TYPES OF ORGANISMS PRESENT IN LAGOONS

Bacteria, *protozoa*, *fungi*, and algae can all be found in stabilization lagoons. The distribution and predominance of various species depends on the chemical and physical characteristics of the wastewater as well as seasonal temperature and ecological *competition*. Algae growth is proportional to light intensity, wastewater temperature, and available nutrients. The amount of available DO significantly affects the type and *dispersion* of organisms within the lagoon. The most important factors that influence lagoon functioning because of their influence on microbial populations are temperature, *pH*, salinity, DO, and light.

BACTERIA. As previously discussed, aerobic, anaerobic, and facultative bacteria will be present in most lagoons. Typical bacterial *morphology* is illustrated in Figure 5.5. Bacteria are responsible for most of the organic waste decomposition. The predominant facultative bacterial *genera* are *Gram-negative*, rod-shaped species: *Alcaligenes, Flavobacterium, Pseudomonas*, and *Achromobacter*. *Coliform bacteria* such as *Escherichia, Enterobacter*, and

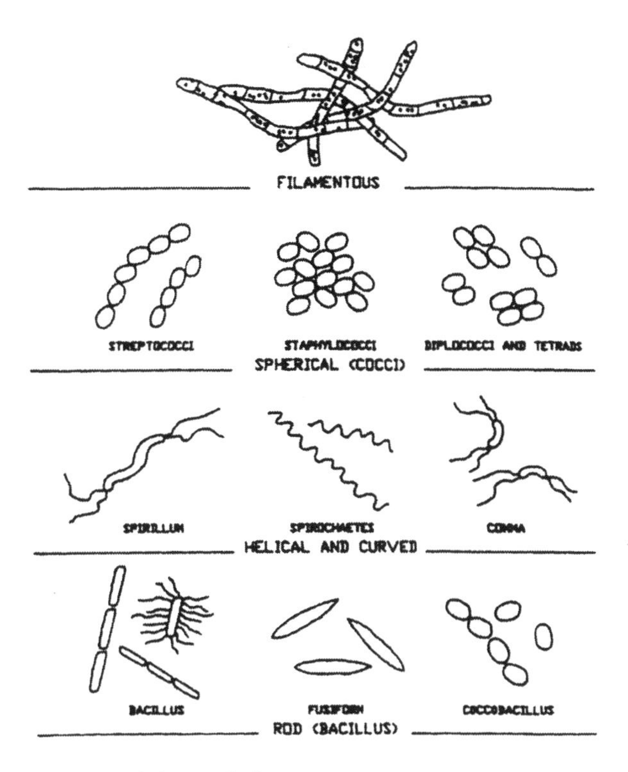

Figure 5.5 Bacterial morphology.

Klebsiella, which predominate in human wastewater, seem to play a minimal role in the organic stabilization process. A high die-off rate of *coliform bacteria* has been observed. *Actinomycetes* (filamentous forms) are also present. The genera *Sphaerotilus, Nocardia,* and *Microthrix* have been observed in aerated lagoons. *Thiothrix* species and other sulfur users may be present under *septic* conditions in lagoons. Filamentous bacteria that may be present include *Sphaerotilus, Haliscomenobacter,* and type 1701. They are indicative of low oxygen concentration and *nutrient deficiency*. In addition, the presence of *Thiothrix, Beggiatoa,* and type 0411 are found in septic conditions. Because of the large amount of algal growth and photosynthetic activity during summer months in temperate climates, the pH is high (above 8.5). This may influence the coliform inactivation rate.

In anaerobic and facultative lagoons, bacteria such as *Methanobacterium, Methanosarcina, Methanococcus,* and *Methanospirillum* will be present in the bottom sludge. They convert organic acids and *hydrolysis* products produced by acid-forming bacteria such as *Alcaligenes* and *Flavobacterium* to methane. Bacteria that are anaerobic, photosynthetic, and hydrogen sulfide (H_2S) oxidizers may develop in facultative lagoons. They use sunlight as energy. Two common genera are *Chlorobium* and *Chromatium*. The latter gives lagoons a characteristic red color in the presence of organic overloading and septic conditions. These bacteria are important in odor control because of their conversion of sulfides.

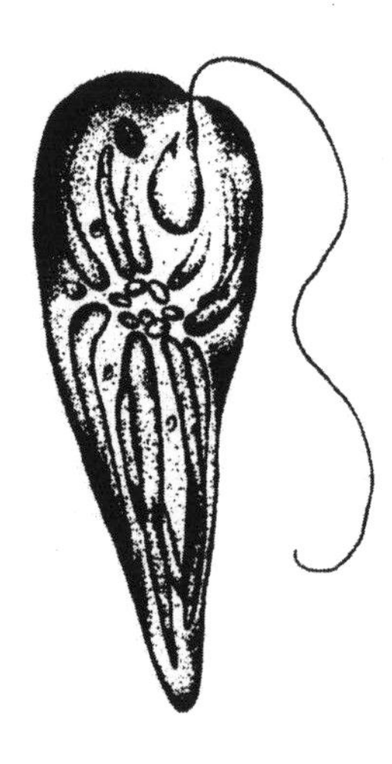

Figure 5.6 *Euglena* **sp.**

PROTOZOA. Protozoan *succession* occurs in stabilization lagoons as it does in other treatment processes. Depending on the environmental parameters of the lagoon, different populations of *protozoa* will be present. Because most protozoa are aerobic, they will be found in the surface zone of facultative lagoons. In aerated lagoons, the protozoan populations will be similar to those found in *activated sludge* and oxidation-ditch processes. The predominant groups of protozoa found in all wastewater treatment processes include the Phytomastigophora, Zoomastigophora, Rhizopoda, Actinopoda, and Ciliophora. In facultative stabilization lagoons, free-swimming *euglenoids* (Phytomastigophora) are a significant part of the population. These species include *Peranema* and *Nostrolenus*, which do not contain *chloroplasts*. The *flagellated* genera *Monas* and *Oicomonas*, which contain golden brown chloroplasts, are also usually represented.

Free-swimming plantlike flagellates with chloroplasts such as the genus *Euglena* (Figure 5.6) are very dominant in lagoons. These types are often classified as single-celled *motile* algae.

Ciliated protozoa are also found in lagoons, although in facultative lagoons they are not as predominant as they are in activated sludge and other aerated treatment processes. The ciliates include the free-swimming genera *Paramecium*, *Cinetochilum*, and *Chilodonella* and the crawling genera *Euplotes* and *Aspidiscus*. Some examples of common Ciliophora are *Lionotus*, *Aspidisca*, *Paramecium*, and *Stylonychia*. These are illustrated in (Figure 5.7).

Sessile stalked ciliates that may be present in aerated lagoons include *Vorticella*, *Carchesium*, *Zoothamnium*, *Opercularia*, and *Epistylis*.

NEMATODES. Free-living wormlike invertebrates, commonly known as thread worms, eel worms, and roundworms (Figure 5.8), may be present in stabilization lagoons. They thrive in aerobic conditions; therefore, in *quiescent* facultative lagoons they are not as readily observed as they are in aerated lagoons. *Nematodes* are *predators* on the microbial populations.

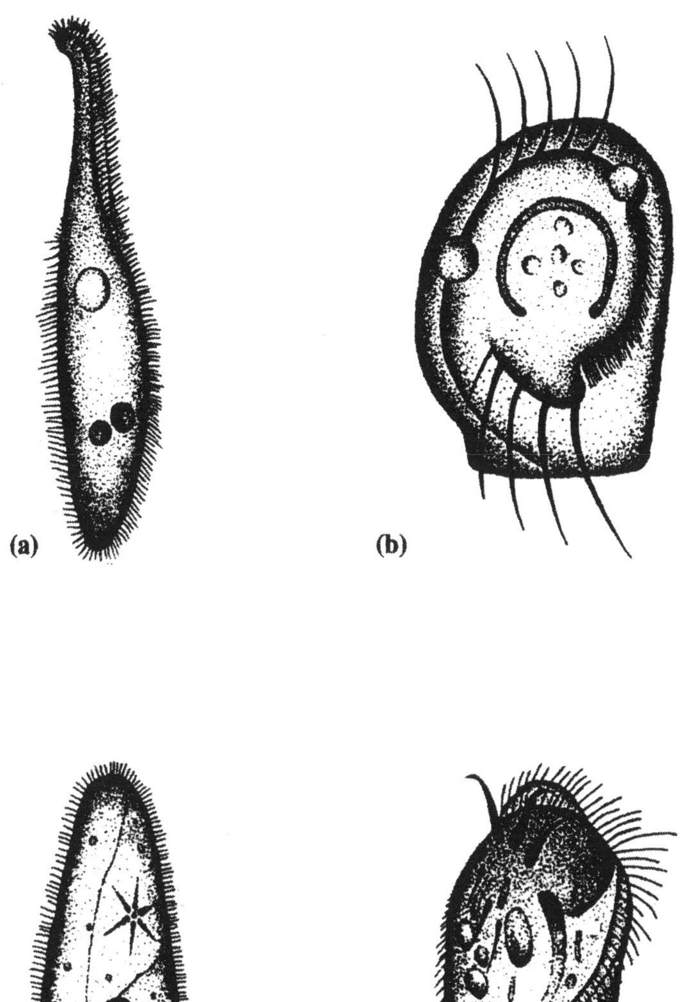

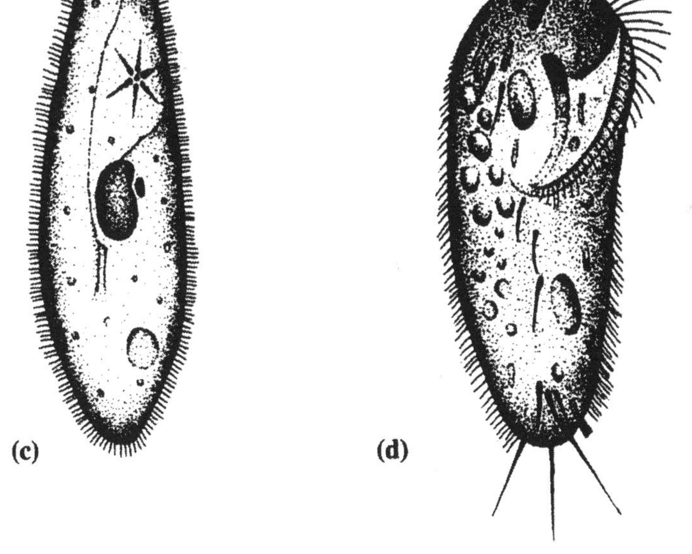

Figure 5.7 Protozoa: (a) *Lionotus,* (b) *Aspidisca*, (c) *Paramecium,* and (d) *Stylonychia*.

Figure 5.8 Roundworms (*Rhabditis*).

ROTIFERS. The *rotifers* are multicellular organisms, which are found in all aquatic systems (Figure 5.9). In aerated lagoons, they will be present under similar circumstances as they are in the activated sludge or oxidation-ditch processes.

ALGAE. The algae found in lagoons are single celled or filamentous, motile or *nonmotile*, plantlike organisms that contain photosynthetic pigments. The algae groups that have been identified in wastewater lagoons include green algae, flagellate algae, *diatoms*, and blue-green algae (presently accepted classification—*cyanobacteria*). Typical examples of genera representative of the four groups found in lagoons include blue-green algae: *Anaebena, Microcystis,* and *Oscillatoria*; diatoms: *Cyclotella, Gomphonema,* and *Achnanthes*; flagellate algae: *Chlamydomonas, Chlorogonium, Phacus, Pyrobotrys,* and *Euglena*; and green algae: *Ankistrodesmus, Chlorella, Scenedesmus, Tetraedron,* and *Anacystis* (Figure 5.10). From previous studies (Palmer, 1977) conducted on

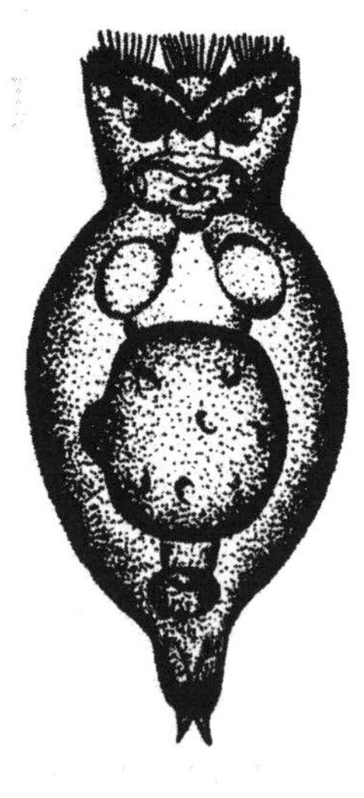

Figure 5.9 Rotifers (*Epiphanes*).

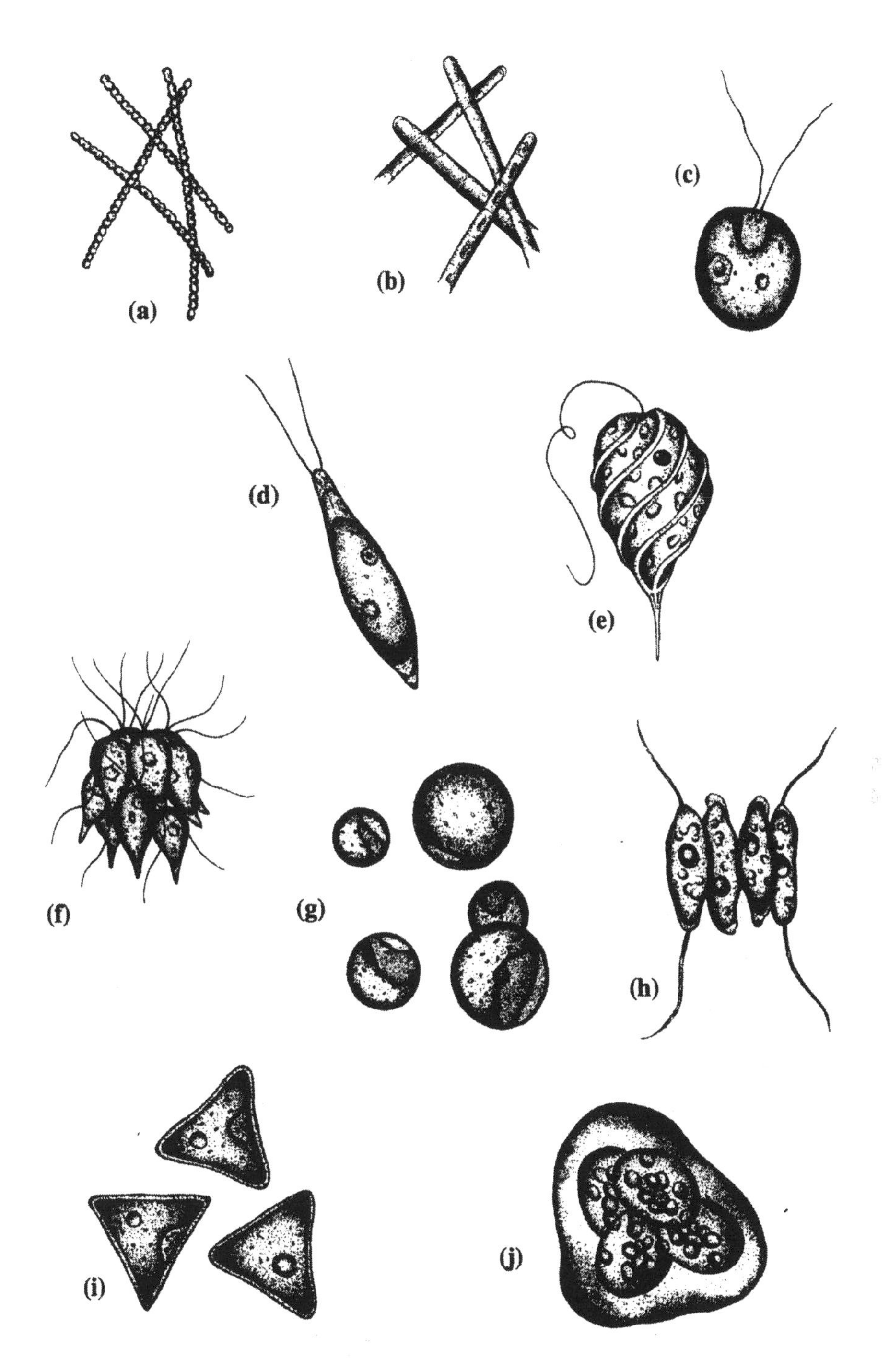

Figure 5.10 Algae: (a) *Anabaena* sp., (b) *Oscillatoria* sp., (c) *Chlamydomonas* sp., (d) *Chlorogonium* sp., (e) *Phacus* sp., (f) *Pyrobotrys* sp., (g) *Chlorella* sp., (h) *Scenedesmus* sp., (i) *Tetraedron* sp., and (j) *Anacystis* sp.

Table 5.1 Common algal genera in U.S. wastewater lagoons (abridged list) (Palmer, 1977).

Achnanthes	*Microspora*
Actinastrum	*Navicula*
Anabaena	*Oedogonium*
Anabaenopsis	*Oocystis*
Anacystis	*Oscillatoria*
Ankistrodesmus	*Palmella*
Aphanizomenon	*Palmellococcus*
Characium	*Pandorina*
Chlamydomonas	*Pediastrum*
Chlorella	*Phacus*
Chlorococcum	*Phormidium*
Chlorogonium	*Pinnularia*
Cladophora	*Pyrobotrys*
Closterium	*Rhodomonas*
Coccomonas	*Selenastrum*
Cryptomonas	*Sphaerocystis*
Cyclotella	*Spirogyra*
Cymbella	*Spirulina*
Desmidium	*Staurastrum*
Dinobryon	*Stichococcus*
Eudorina	*Synedra*
Euglena	*Tetradesmus*
Fragilaria	*Ulothrix*
Gomphonema	*Zygnema*
Lyngbya	

algae populations in wastewater lagoons across the United States, 50% were green algae, 25% were pigmented flagellates, 15% were blue-green algae, and 10% were diatoms.

Algae populations in wastewater lagoons vary according to chemical and physical lagoon characteristics, including temperature, pH, and nutrients. *Chlorella, Ankistrodesmus, Scenedesmus, Euglena, Chlamydomonas, Oscillatoria, Micratinium, Anacystis*, and *Oocystis* have been found to be dominant in decreasing order of abundance throughout lagoons in the United States, although 125 different genera have been recorded (Table 5.1).

HIGHER LIFE FORMS

Many higher life forms develop in lagoons. These include protozoa and multicellular organisms such as rotifers, cladocerans (*Daphnia*) (Figure 5.11), chironomids (midges) (Figure 5.12), mosquito *larvae* and adults (Figure 5.13)

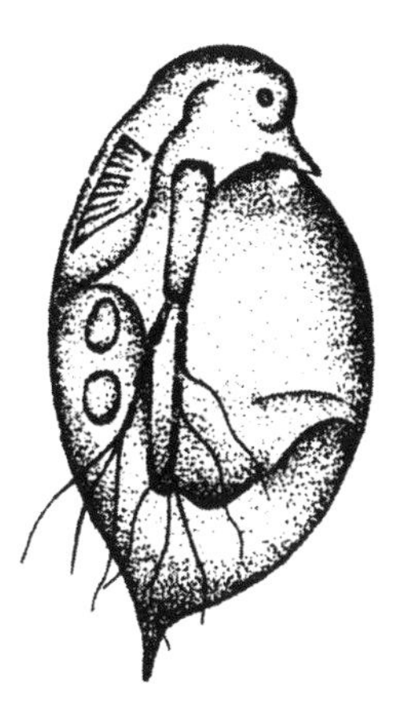

Figure 5.11 Crustaceans (*Daphnia*).

and *Psychoda* (flies) (Figure 5.14). These organisms play a role in wastewater treatment by feeding on algae and bacteria and promoting *flocculation* and settling of particulate matter. Aquatic beetles such as the whirligig beetle larvae and adult water *scavenger* beetles are also common (Figure 5.15). Mosquitoes grow in some lagoons and may cause a nuisance and public health problem. *Culex tarsalis*, a vector of *encephalitis*, may grow in lagoons. *Daphnia* blooms occur in the summer in temperate climate lagoons. Some multiple-cell stabilization lagoons are designed to include *aquaculture*, that is, the use of water plants such as the *water hyacinth* and various *species* of fish, to provide enhanced treatment and commercial products.

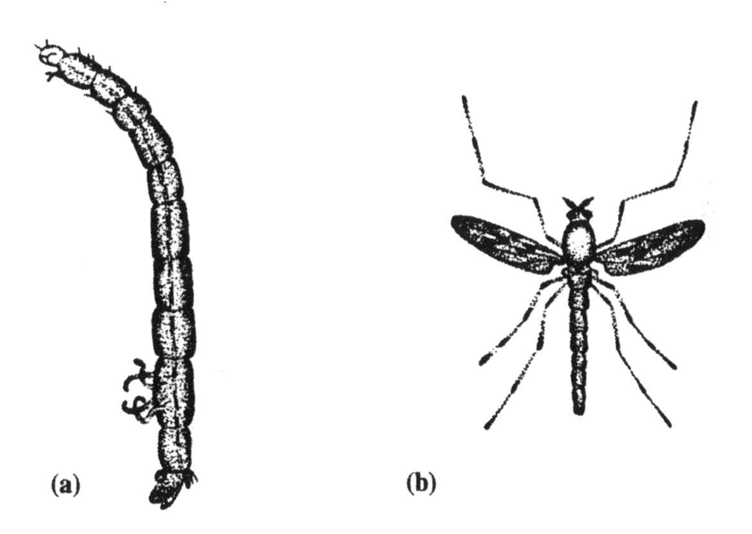

Figure 5.12 Chironomids (midges): (a) larva (*Chironomus*) and (b) adult (*Chironomus*).

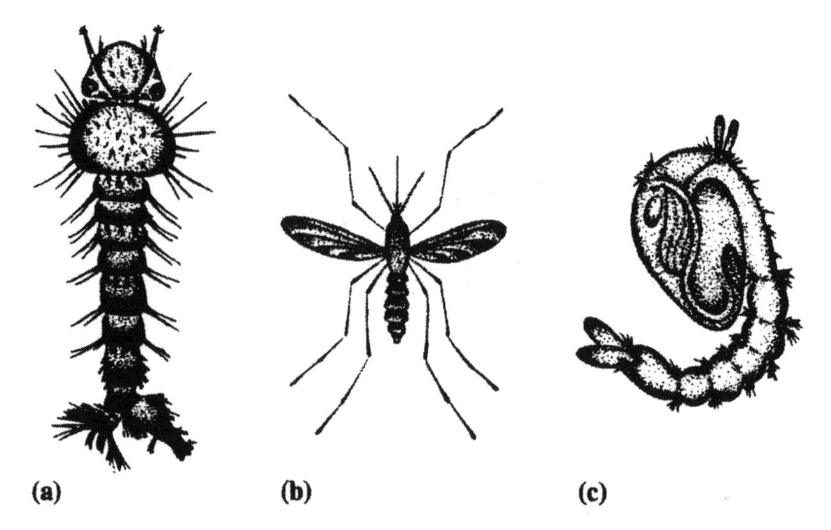

Figure 5.13 Mosquitoes: (a) larva (*Aedes*), (b) adult (*Aedes*), and (c) pupa (*Culex*).

OPERATION OF STABILIZATION LAGOONS

FACULTATIVE LAGOONS. Facultative lagoons, which are usually 1.2 to 2.4 m (4 to 8 ft) deep, have an upper aerobic layer overlying an anaerobic layer, which illustrates a vertical zonation (Figure 5.16). The aerobic layer is

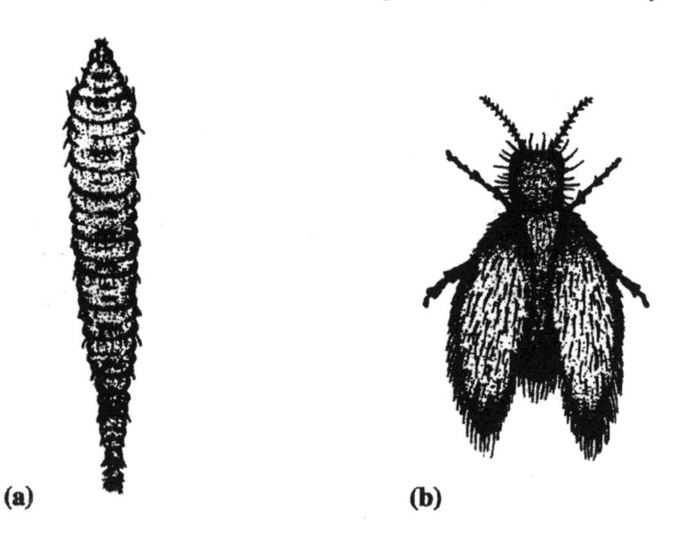

Figure 5.14 Wastewater fly: (a) larva (*Psychoda*) and (b) adult (*Psychoda*).

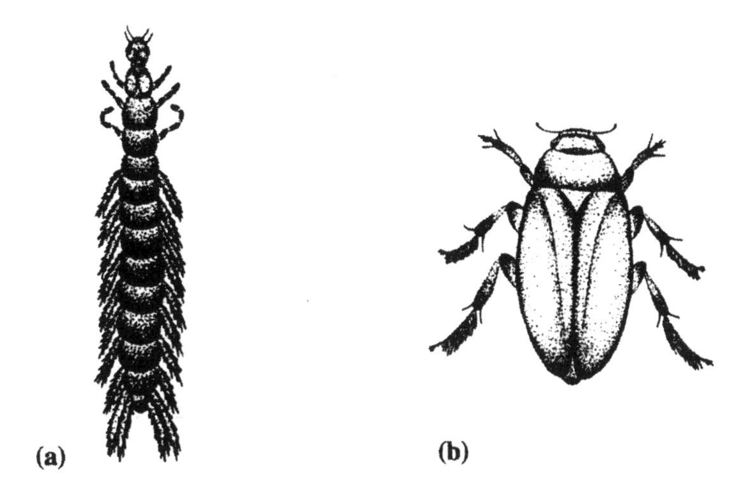

Figure 5.15 Aquatic beetles: (a) larva whirligig beetle (*Dineutus*) and (b) adult water scavenger beetle (*Hydrophilus*).

important in maintaining an oxidizing environment in which gases escaping the lower anaerobic layer are oxidized, preventing odors. Natural surface *aeration* from wind mixing and algal photosynthesis supplies the oxygen needed for aerobic bacterial stabilization of the wastewater. *Detention times* vary from 5 to 30 days in temperate climates.

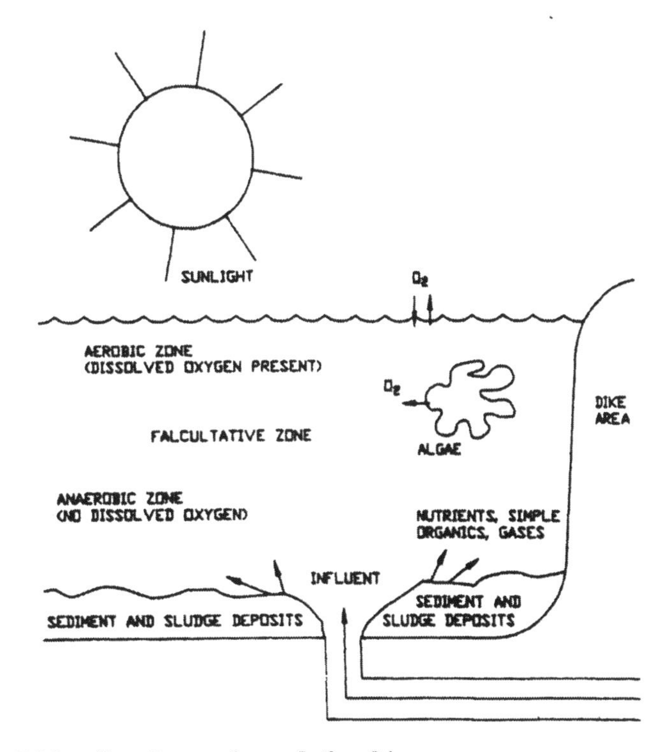

Figure 5.16 Zonal organism relationship.

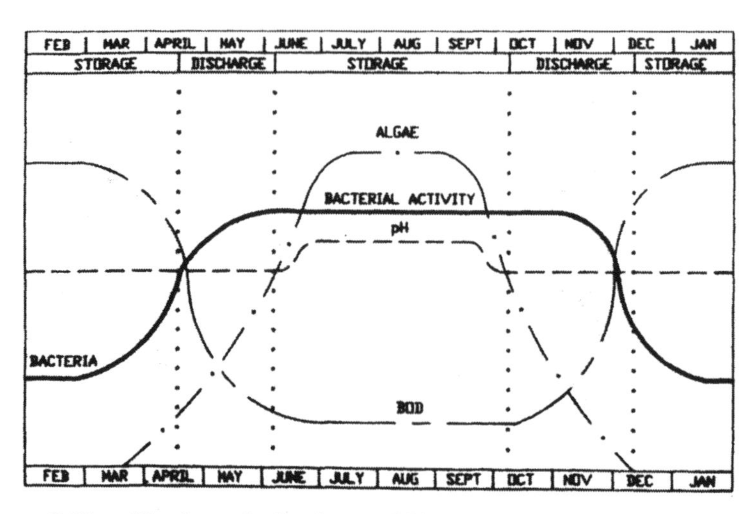

Figure 5.17 Yearly variation in a stabilization lagoon.

Aerobic stabilization of BOD by aerobic bacteria occurs in the upper oxygenated layer. Oxygen is partly supplied by algal photosynthesis, while BOD is converted to methane in the lower anaerobic layer by methane bacteria. There is a symbiosis between bacteria and algae in this process. A balance of microbial populations is necessary for successful operation.

Diurnal variations involving DO concentrations in a facultative lagoon occur over a 24-hour period. Dissolved oxygen concentration can reach saturation at midafternoon because of algal photosynthesis. At night, DO concentration drops to a minimum from respiration in the absence of photosynthesis. Supplemental mechanical aeration may be used at night to prevent odor production resulting from anaerobic bacterial activity.

Facultative lagoons are easy treatment systems to operate, but performance limiting factors include the need for a large surface area to prevent organic overloading. Expected effluent values range from 20 to 60 mg/L BOD and 30 to 200 mg/L SS. Seasonal variations in stabilization lagoon efficiency occur in temperate climates because of temperature variations. Temperature influences the growth and metabolic activity of lagoon organisms. When the temperature falls to lower than 4 °C, little stabilization occurs and the lagoon becomes a storage facility. Optimum performance occurs in summer within temperatures of 20 to 35 °C. Therefore, organic loading rates have to be reduced in colder climates. Most conventional lagoons that are subject to winter ice cover are now operated on a seasonal discharge basis (Figure 5.17).

Algae blooms can cause SS removal problems in facultative lagoons. This also causes increased *alkalinity* in a lagoon (pH of 9 to 10) because of the consumption of carbon dioxide (CO_2) during photosynthesis. Three methods are used to remove algae from lagoon effluents: land treatment, sand filtration, and chemical addition of a *coagulant* such as alum to precipitate the algal solids.

AERATED LAGOONS (MECHANICALLY ASSISTED LAGOONS). The principal source of oxygen for an aerated lagoon is mechanical agitation.

Aeration can be provided by brush aerators, propeller pumps, or a diffused air system. Many facultative lagoons have been converted to aeration to solve overloading problems. Aerated lagoons require less land area than facultative ones, but operational costs are higher.

Continuous mixing is not conducive to algae growth. If photosynthesis is desirable, intermittent mixing should be used. Aerated lagoons often operate like activated sludge processes or oxidation ditches with less biomass. Activated sludge processes involve *biomass* increases with recycling of solids.

PERFORMANCE OF LAGOONS

Lagoons remove BOD and TSS by bacterial oxidation to CO_2 and through settling. Biomass, which is created during treatment, settles to become lagoon sludge. Settled organic materials may undergo further biodegradation through anaerobic activity.

Lagoons typically remove 85 to 95% of the influent BOD and TSS and optimally may produce an effluent of <30 mg/L BOD and TSS. However, "new" BOD and TSS are produced during treatment as algal biomass. Algae are part of the BOD and TSS and may represent a significant contribution to effluent BOD and TSS. Algal biomass in a lagoon can raise effluent BOD to 60 to 100 mg/L and effluent TSS to 60 to 150 mg/L. Lagoon effluent BOD and TSS concentrations may significantly be the result of algae growth and not poor influent wastewater treatment. The diagnosis of an algae overgrowth problem rather than a basic wastewater treatment problem is important.

Nitrogen in wastewater is principally in the form of organic nitrogen (*protein*) and ammonia. Organic nitrogen is converted to ammonia by ammonifying bacteria in the lagoon. Three processes remove ammonia-nitrogen: (1) stripping to the atmosphere, significant at pH values greater than pH 8, often induced by algal growth; (2) *assimilation* to organisms (bacteria and algae); and (3) bacterial *nitrification* followed by *denitrification*.

Approximately 50 to 90% of influent nitrogen may be removed in lagoon processes. Bacterial biomass produced during wastewater treatment contains approximately 10% nitrogen by weight. Because most lagoons produce 50 to 100 mg/L bacterial biomass for an influent BOD concentration of 200 mg/L (a 25 to 50% yield), only 5 to 10 mg/L of ammonia is removed by bacterial cell growth. Additional ammonia can be removed by bacterial nitrification to nitrate (NO_3^-) and may also be subsequently denitrified to nitrogen gas (N_2). Bacterial nitrification and denitrification are variable in lagoons but can produce an effluent with <1 mg/L ammonia and total nitrogen. Nitrification is strongly temperature dependent and ceases below a temperature of approximately 5 °C. Ammonia removal in lagoons is strongly dependent on season. Complete removal in the summertime may be affected in warmer climates but ceases at lower temperatures in the wintertime in colder climates.

Organic loading rates have to be reduced in colder climates during the winter months, and *hydraulic detention times* need to be increased.

Reference

Palmer, C.M. (1977) *Algae and Water Pollution.* EPA-600/90-77-036, U.S. EPA, Cincinnati, Ohio.

Suggested Readings

American Public Health Association, American Water Works Association, and Water Environment Federation (1995) *Standard Methods for the Examination of Water and Wastewater.* 19th Ed., Washington, D.C.

Fitzgerald, G.P. (1964) The Effect of Algae on BOD Measurements. *J. Water Pollut. Control Fed.*, **36,** 1524.

Metcalf & Eddy, Inc. (1991) *Wastewater Engineering: Treatment, Disposal, and Reuse.* 3rd Ed., McGraw–Hill, New York.

Patterson, D.J., and Hedley, S. (1992) *Free-Living Freshwater Protozoa—A Color Guide.* CRC Press, Boca Raton, Fla.

Pennak, R.W. (1989) *Fresh-Water Invertebrates of the United States, Protozoa to Mollusca.* 3rd Ed., Wiley & Sons, New York.

Prescott, G.W. (1978) *How To Know the Freshwater Algae.* 3rd Ed., Brown, Dubuque, Iowa.

U.S. Environmental Protection Agency (1983) *Municipal Wastewater Stabilization Lagoons Design Manual.* EPA-625-1-83-015, U.S. EPA, Center Environ. Res. Inf., Cincinnati, Ohio.

Water Environment Federation (1990) *Wastewater Biology: The Microlife.* Special Publication, Alexandria, Va.

Water Environment Federation (1994) *Wastewater Biology: The Life Processes.* Special Publication, Alexandria, Va.

Whitford, L.A., and Shumacher, G.J. (1984) *A Manual of Fresh-Water Algae.* Sparks Press, Raleigh, N.C.

Chapter 6
Wetlands

INTRODUCTION

Wetlands are land areas that are wet during all or part of the year. Frequently, wetlands are transitional between uplands (terrestrial areas) and persistent or deeply (greater than 2 m) flooded systems (e.g., streams, rivers, deep ponds, and lakes). Wetlands have been referred to by a host of terms, including *marsh*, wet meadow, *bog*, swamp, and *bottomland* forests. Smith (1996) categorizes 20 separate types of freshwater and saline wetlands. The unifying principle of a wetlands is that it is wet long enough to alter soil properties because of chemical, physical, and biological changes that occur during flooding. These changes exclude plant *species* that cannot grow in wet soils.

From an ecological standpoint, there are three key attributes of wetlands: (1) *hydrology* or the degree of flooding or soil saturation; (2) wetlands vegetation or the presence of water-loving (*hydrophytic*) plants; and (3) *hydric* soils or soils that are saturated, flooded, or ponded long enough to develop *anaerobic* conditions in the upper part. All areas considered wetlands must have enough water at some time during the growing season to stress plants and animals not adapted for life in water or saturated soils. Although a number of definitions exist (Mitsch and Gosselink, 1993), most wetlands typically fall within one of the following four categories: (1) areas with both hydrophytes and hydric soils (e.g., marshes, swamps, and bogs), (2) areas without hydrophytes but with hydric soils (e.g., tidal flats), (3) areas without hydric soils but with hydrophytes (e.g., seaweed or covered rocky shores), and (4) periodically flooded areas without hydric soil and without hydrophytes (e.g., gravel beaches or stream beds).

Wetlands *ecosystems* have an intrinsic ability to modify or trap a wide spectrum of waterborne pollutants or contaminants. Although the concept of deliberately using wetlands for wastewater treatment has only developed within the past 40 years, in reality human societies have indirectly used natural wetlands for waste management for thousands of years. Observations of the water purification phenomenon of natural wetlands systems have stimulated development of constructed wetlands for wastewater treatment. Constructed wetlands, in contrast to natural wetlands, are designed, built, and operated to emulate wetlands or functions of natural wetlands and have recently received considerable attention as a low-cost, efficient means to clean up many types of wastewater. Natural or constructed wetlands have been used and/or designed to treat not only municipal wastes but also point and nonpoint wastes such as acid mine drainage, agricultural wastes, landfill *leachate*, paper and pulp, petrochemicals, and industrial wastes. The use of wetlands for treatment of wastewater is an emerging technology in North America and worldwide (Moshiri, 1993, and Kadlec and Knight, 1996).

HOW WETLANDS TREAT WASTEWATER

Wetlands accomplish wastewater treatment through a variety of physical, chemical, and biological processes operating independently in some circumstances and interacting in others (see Table 6.1). Wetlands have a higher rate of biological activity than most ecosystems; they can transform many of the common pollutants that occur in conventional wastewater to harmless products or essential nutrients that can be used for additional biological productivity. These transformations are accomplished by virtue of the wetlands' land area, with its inherent natural environmental energies of sun, wind, soil, plants, and animals (Kadlec and Knight, 1996).

The mechanisms in wetlands ecosystems that modify dissolved and *particulate* substances in wastewater include *sedimentation, adsorption,* filtration, *precipitation, volatization, complexing, microbial* modification, and vegetation uptake (Kadlec, 1989; Majumdar et al., 1989; and Kadlec and Knight, 1996). Sedimentation may be the most significant initial process because suspended solids have a strong tendency to adsorb *pathogenic organisms, refractory organics, hydrocarbons, heavy metals,* and nutrients. Vegetation obstructing the flow and reducing the velocity enhances sedimentation. Filtration, precipitation, and complexing are interrelated and, to a large extent, dependent on hydraulic resistance from vegetation and soils that enhance sedimentation. The increased retention time that wetlands provide

Table 6.1 Contaminant removal processes in wetlands.[a]

Process	Contaminant
Biological	
Microbial metabolism	Colloidal and settleable solids, organics, nutrients, metals, biodegradable organics (BOD_5,[b] COD[c])
Predation	Pathogens
Natural die-off	Pathogens
Plant metabolism	Refractory organics, pathogens
	Nutrients, metals, refactory organics
Chemical	
Precipitation	Metals, salts
Adsorption	Metals, salts
Oxidation–reduction	Refractory organics, metals
Volatilization	Refactory organics, nutrients
Physical	
Sedimentation	Colloidal and settleable solids (includes metals, biodegradable organics, pathogens, refactory organics, and nutrients)
Filtration	Colloidal and settleable solids
Adsorption	Coloidal solids
Photolysis	Refractory organics, pathogens

[a]Adapted from Hammer (1995).
[b]BOD = biochemical oxygen demand.
[c]COD = chemical oxygen demand.

improve secondary treatment via *oxidation* and volatilization of many substances.

Microbial populations in the water column, attached to vegetative and other *substrates* and within soils, modify hydrocarbons, metals, *pathogens*, and nutrient loads, causing precipitation of some pollutants and recycling with subsequent settling of others. Not only does a dense stand of wetlands vegetation provide a large surface area within the water column for microbial attachment, but the natural oxygen loss from wetlands plant root structures creates a substantial *aerobic* environment for microbial populations within the soil.

Hydrophytic, or wet-growing plants, have specialized structures in their leaves, stems and roots somewhat analogous to a mass of breathing tubes (i.e., "snorkels") that conduct oxygen down into the roots. Because the outer covering on the root hairs is not a perfect seal, oxygen leaks out, creating a thin film aerobic region (the *rhizosphere*) around every root hair. The larger region outside the rhizosphere remains anaerobic, but the positioning of a large thin film aerobic region surrounded by an anaerobic region is crucial to the growth of microbial populations involved in the transformation of nitrogenous compounds and other substances.

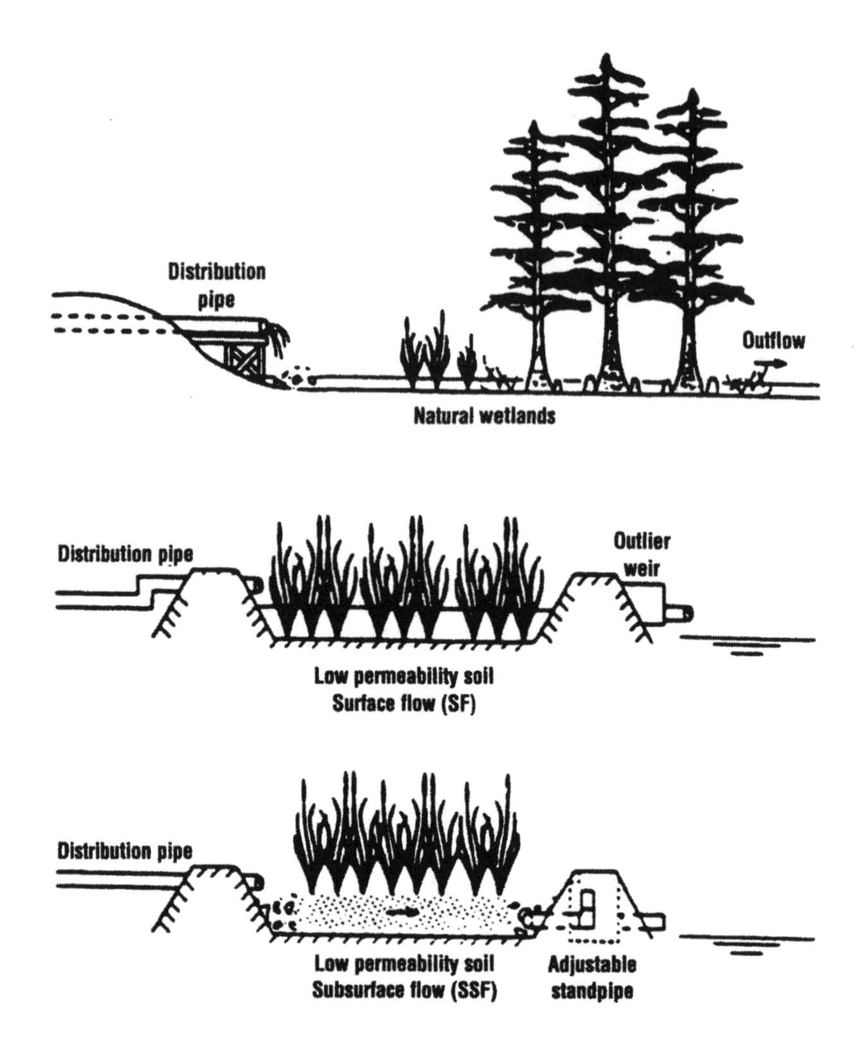

Figure 6.1 Three basic wetlands treatment system types (adapted from Kadlec and Knight, 1996).

TYPES OF WETLANDS TREATMENT SYSTEMS

Wetlands treatment systems (natural and constructed) use rooted, water-tolerant plant species and shallow, flooded, or saturated soil conditions to provide various types of wastewater treatment (Hammer, 1995). The three basic types of wetlands treatment systems include natural wetlands (NWs), constructed surface-flow wetlands, and constructed sub-surface-flow (SSF) wetlands (Figure 6.1).

Although there are many types of naturally occurring wetlands, only those types with plant species that are adapted to continuous flooding are suitable for receiving continuous flows of wastewater. Also, because of their protected regulatory status, discharges to NWs must receive a high level of pretreatment (minimum of secondary). Constructed wetlands mimic the optional treatment conditions found in NWs and provide the flexibility of being built in almost any location. In addition, constructed wetlands can be used for pretreatment of primary and secondary wastewater from a variety of sources (Kadlec and Knight, 1996).

Surface-flow wetlands (natural and constructed) are densely vegetated by a variety of plant species and typically have water depths less than 0.4 m. Open water areas may be incorporated to the design to provide for optimization of hydraulics and wildlife *habitat*. Sub-surface-flow wetlands use a bed of soil or gravel as a substrate for the growth of rooted plants. Pretreated wastewater flows by gravity horizontally through a bed substrate where it contacts a mixture of facultative microbes living in association with the substrate and plant roots. Bed depth in SSF wetlands is typically less than 0.6 m, and the bottom of the bed is sloped to minimize the water flow overland. Typical plant species in SSF wetlands include reed (*Phragmites*), cattail (*Typha*), and bulrush (*Scirpus*).

CONDITIONS INFLUENCING MICROBIAL POPULATIONS

The diverse microbial populations in wetlands include *bacteria*, *fungi*, and *algae*, which are important for nutrient cycling and pollutant transformations. The variety of microbial species in wetlands function in a wide range of physical and chemical conditions. Because of this variety of species and the *niches* they occupy, wetlands ecosystems can operate consistently to treat wastewater. Because many of these organizations are the same as those important in conventional treatment systems (see Table 6.2), their growth requirements and characteristics are known (Kadlec and Knight, 1996).

Overall ecosystem parameters such as *dissolved oxygen*, water temperature, and *influent* constituent concentrations must be controlled through design and system operation to keep the *microbial community* in harmony for optimal treatment. Wetlands vegetation substantially increases the amount of environment (aerobic and anaerobic) available for microbial populations both aboveground and belowground. There is no doubt that *macrophytic* plants are essential for the high-quality treatment performance of most wetlands systems. The numerous studies (Kadlec and Knight, 1996) measuring treatment with and without plants have concluded that, almost invariably, performance is greater when plants are present. Because of this finding, wetlands plants were once believed to be the dominant source of treatment because of their direct uptake and sequestering of pollutants. It is now known that direct uptake is the

Table 6.2 Representative organisms important in wetlands treatment systems.[a]

Group	Representative genera
Bacteria	
Phototrophic	*Rhodospirillum* (nonsymbiotic nitrogen fixer)
Gliding	*Beggiatoa* (oxidizes hyrdogen sulfide)
	Flexibacter, Thiothrix
Sheathed	*Sphaerotilus* (common in wastewater treatment plants)
Gram-negative	*Escherichia* (predominant coliform in feces)
	Enterobacter (nonsymbiotic nitrogen fixer)
	Desulfovibrio (reduces sulfate to hydrogen sulfide)
	Nitrosomonas, Nitrobacter (reduction of NH_4^{+}[b] to NO_3^-)
Methane producing	*Methanobacteruim* (anaerobic, converts carbonate to methane)
Gram-positive	*Streptococcus*
Endospore forming	*Clostridium, Bacillus*
Actinomycetes	*Nocardia, Frankia*
Fungi	*Cryptococcus, Dactyella* (yeasts)
Algae	
Cyanophyta (blue-green)	*Anabaena, Oscillatoria*
Chlorophyta (green)	*Chlorella, Chlamydomonas*
Euglenophyba (euglenoids)	*Euglena, Phacus*
Pyrrophyta (dinoflagellates)	*Gymnodinium, Peridinium*
Chrysophyta (yellow-green)	*Chromulina, Dinobryon*
Bascillariophyta (diatoms)	*Navicula, Fragillaria*
Cryptophyta (crytomonads)	*Cryptomonus*
Rhodophyta (red)	*Bangia*
Phaeophyta (brown)	*Fucus*
Macrophytes	
Mosses	*Fontinalis*
Ferns	*Woodwardia*
Conifers	*Pinus*
Monocots	*Typha, Lemna, Wolffia, Hydrilla, Phragmites, Spartina, Cyperus, Sagittaria*
Dicots	*Acer, Betula, Populus, Salix*

[a]Adapted from Kadlec and Knight (1996).
[b]NH_4^+ = ammonium.

principal removal mechanism only for some pollutants and only in lightly loaded systems. During an initial successional period of rapid plant growth, direct pollutant immobilization in wetlands plants may be important. For many other pollutants, plant uptake is generally of minor importance compared to microbial and physical transformations that occur in wetlands.

Macrophytic plants are essential in wetlands treatment systems because they provide structure for the microbes that mediate most of the pollutant transformations. Any direct uptake of nutrients, metals, and *organic* substances by plants provides some removal for a limited time but litter decomposition usually releases these compounds back to the water column after the growing season. Consequently, the most important role of plants is to simply grow up and die (creating *detritus*), and the water quality improvement function is principally dependent on the high conductivity of the litter/*humus* (detritus) layer and the large area for microbial attachment. Sedimentation and subsequent microbial modification (*metalization*, *oxidation*, uptake, etc.) are likely the most important processes for direct removal of pollutants from wastewater to soil substrate "sinks" inaccessible to future inflows and flushing actions (Hammer, 1995).

In summary, wetlands systems use larger land areas and natural energy inputs to establish self-maintaining treatment systems, providing environments for many more types of organisms because of the diversity of microenvironments in a wetlands. The latter, along with a large treatment area, frequently provide more complete reduction and lower discharge concentrations of waterborne contaminants (Gillette, 1996, and Kadlec and Knight, 1996). More detailed information about wetlands microbial communities (bacteria, fungi, and algae) can be found in Ainsworth et al. (1973), Kadlec and Knight (1996), Portier and Palmer (1989), and South and Whittick (1987). Information about the ecology of vascular plant species found in wetlands can be found in Kadlec and Knight (1996) and Mitsch and Gosselink (1993).

References

Ainsworth, G.C.; Sparrow, F.K.; and Sussman, A.S. (Eds.) (1973) *The Fungi: An Advanced Treatise*. Academic Press, New York.

Gillette, B. (1996) Wetlands Treatment: Getting Close to Nature. *BioCycle*, **37**, 74.

Hammer, D.A. (1995) Water Quality Improvement Functions of Wetlands. In *Encyclopedia of Environmental Biology*. W.A. Nierenberg (Ed.), Academic Press, New York, 667.

Kadlec, R.H. (1989) Wetlands for Treatment of Municipal Wastewater. In *Wetlands Ecology and Conservation: Emphasis in Pennsylvania*. S.K. Majumdar, R.P. Brooks, F.J. Brenner, and R.W. Tiner, Jr. (Eds.), The Pennsylvania Academy of Science, Easton, Pa., 395.

Kadlec, R.H., and Knight, R.L. (1996) *Treatment Wetlands*. Lewis Publishers, New York, 893.

Majumdar, S.K.; Brooks, R.P.; Brenner, F.J.; and Tiner, R.W. (Eds.) (1989) *Wetlands Ecology and Conservation: Emphasis in Pennsylvania.* The Pennsylvania Academy of Science, Easton, Pa.

Mitsch, W.J., and Gosselink, J.G. (1993) *Constructed Wetlands for Quality Improvement.* Lewis Publishers, New York, 535.

Moshiri, G.A. (Ed.) (1993) *Constructed Wetlands for Water Quality Improvement.* Lewis Publishers, Ann Arbor, Mich.

Portier, R.J., and Palmer, S.J. (1989) Wetlands Microbiology: Form, Function, Processes. In *Constructed Wetlands for Wastewater Treatment: Municipal, Industrial, and Agricultural.* D.A. Hammer (Ed.), Lewis Publishers, Chelsea, Mich.

Smith, R.L. (1996) *Ecology and Field Biology.* Harper Collins, New York, 740.

South, G.R., and Whittick, A. (1987) *Introduction to Phycology.* Blackwell Science, Oxford, U.K.

SUGGESTED READINGS

Cooper, P.F., and Green, M.B. (1994) Reed Bed Treatment Systems for Sewage Treatment in the United Kingdom—The First Ten Years Experience. *Proc. 4th Int. Conf. Wetland Syst. Water Pollut. Control,* Guangzhou, China, 565.

Lyons, J.G. (1993) *Practical Handbook for Wetlands Identification and Delineation.* Lewis Publishers, Ann Arbor, Mich., 157.

Miller, G.T. (1997) *Environmental Science—Working with the Earth.* Wadsworth Publishing, New York, 519.

Schierup, H.H.; Brix, H.; and Lorenzen, B. (1990) Wastewater Treatment in Constructed Reed Beds in Denmark—State of the Art. In *Proceedings of the International Conference on the Use of Constructed Wetlands in Water Pollution Control.* P.F. Cooper and B.C. Findlater (Eds.), Pergamon Press, Oxford, U.K.

Tennessen, K.J. (1993) Production and Suppression of Mosquitoes in Constructed Wetlands. In *Constructed Wetlands for Water Quality Improvement.* G.A. Moshiri (Ed.), Lewis Publishers, Ann Arbor, Mich., 157.

Water Pollution Control Federation (1990) *Natural Systems for Wastewater Treatment.* Manual of Practice No. FD-16, Alexandria, Va.

Chapter 7
Composting

INTRODUCTION

Composting is an *aerobic microbial* process resulting in decomposition and *stabilization* of *organic substrates.* During composting, microbial *metabolic* activity produces an increase in temperature. The end product is an organic soil conditioner that has been stabilized to a *humus*like product that is free of viable human and plant *pathogens* and weed seeds, does not attract insects or vectors, can be handled and stored without odor being produced, and is beneficial to the growth of plants (Haug, 1993).

The primary purposes of composting wastewater solids are to kill pathogens and to make the biosolids aesthetically acceptable so that they may be distributed safely in public areas. A secondary purpose of some composting systems is to dry the solids from an incoming moisture content of 75 to 80% to a product moisture content of approximately 40%.

STAGES IN COMPOSTING

A composting system consists of several stages (Composting Council, 1994), including preprocessing, high-rate phase, stabilization phase, curing, and postprocessing.

PREPROCESSING. Dewatered cake is too wet (75 to 85% water) and lacks the necessary support structure to provide the free air space needed to maintain an aerobic environment. During preprocessing, a *bulking* agent is added to reduce the unit moisture content to approximately 60% and provide free air space. A good bulking agent must have sufficient porosity and structural integrity and must be sufficiently dry. Wood chips or sawdust are most commonly used as bulking agents. Typical ratios of bulking agent to solids range from 1:1 to 2.5:1 to give a final solids concentration of approximately 40%.

Aerobic microbial activities are most vigorous on the outer surfaces of sludge particles in the mix. Particle size is, therefore, a key factor in preparing the compost mix.

HIGH-RATE AND STABILIZATION PHASES. *Mesophilic* organisms, those that grow only at temperatures below 45 °C, initiate the high-rate phase by rapidly metabolizing the readily degradable components of the wastewater solids. Their metabolic activity results in a rise in temperature (Figure 7.1) and, as it rises above 45 °C, there is a shift of metabolic activity from the mesophilic to *thermophilic* organisms, those that grow at temperatures higher than 45 °C. This shift is often detectable by continuously monitoring the rate

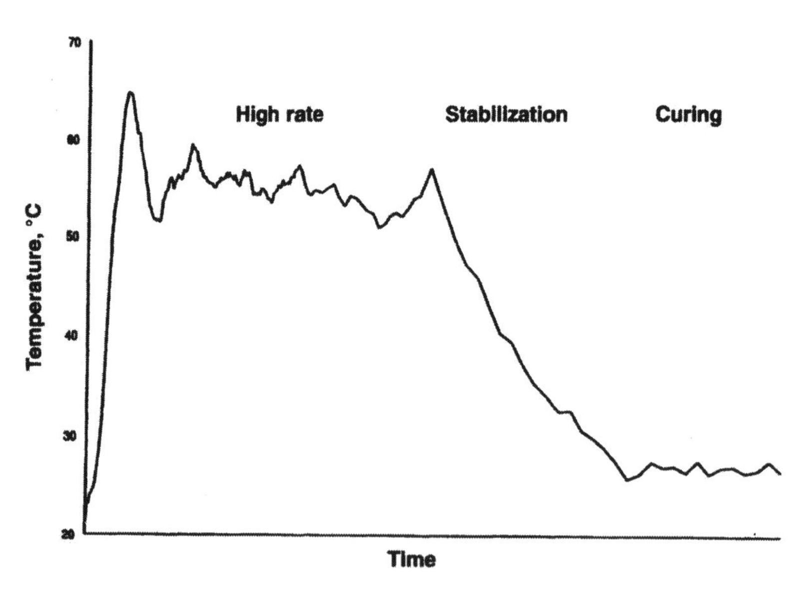

Figure 7.1 Temperature profile ([1.800 × °C] + 32 = °F).

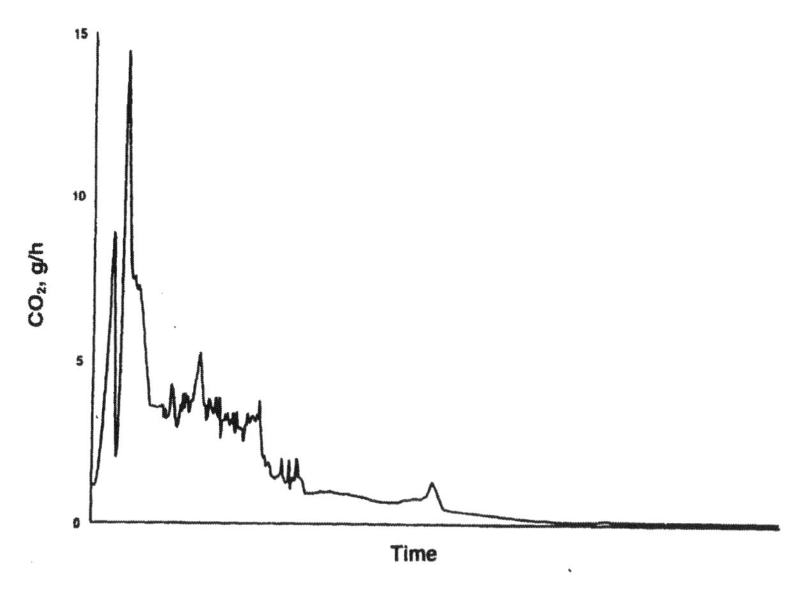

Figure 7.2 Rate of carbon dioxide production.

of oxygen uptake or carbon dioxide production during composting (Figure 7.2). Two sharp peaks of metabolic activity occur as the temperature rises, the first because of the mesophiles and the second because of the thermophiles.

After the easily metabolizable components are degraded, microbial populations continue to decompose the more slowly degradable nutrients, for example, *cellulose* and other *polymers*. When the rate of *metabolism* is no longer sufficient to counterbalance heat loss from the compost mix, the temperature begins to decline. This phase of reduced metabolic rate and decreasing temperature is referred to as the stabilization phase (Figure 7.1).

CURING. After the temperature decreases to near ambient, the mix is transferred to piles to complete the composting process. During the high-rate and stabilization phases, substances such as *acetic acid* that inhibit seed germination and plant growth are produced (Composting Council, 1994). In the curing phase, these *phytotoxic* compounds are degraded. Consequently, completion of the curing phase is best determined by seed germination and plant growth tests. Slowly degradable compounds such as cellulose, *lignin*, *proteins*, *lipids*, and *carbohydrates* continue to be degraded, or they are converted by organisms to *humic substances*. Humic substances are a complex group of chemicals derived from plant and microbial structures that make up the organic content of soil and give it a rich, dark color. In soil, humic substances exist in a dynamic state, being formed when materials decay and providing organisms with a long-term source of nutrients when more easily degradable compounds are not available (Alexander, 1977).

To achieve optimum curing in the minimum amount of time, moisture should be adjusted to approximately 45 to 50% when the curing piles are formed. *Leachate* water should not be used to moisturize stabilized material because it can lead to odor production and reintroduce pathogens. The upper temperature limit for curing is approximately 50 °C. Temperature and sufficient *aeration* can be maintained by air blowers and/or turning. The turning and mixing of piles once a week to once a month during curing breaks up clumps to expose new surfaces and redistributes moisture.

Microbial growth during the curing process may pose an occupational safety risk for certain persons. The fungus, *Aspergillus fumigatus,* tends to grow on cellulosic *substrates* at the reduced curing temperatures, and persons with conditions such as *asthma*, *tuberculosis*, or *diabetes* are more susceptible to respiratory infection by this organism than those without predisposing conditions. Affected persons should be excluded from working near curing processes.

POSTPROCESSING. Compost produced from wastewater solids is generally clean in that it is not necessary to remove glass, metal, plastics or other objects found in municipal solid waste or other composts. Screening may be desirable, however, to recover the bulking agent. Depending on the future use of the compost, it may be necessary to adjust the moisture and test for pathogens, nutrients, and phytotoxic substances.

COMPOSTING TECHNOLOGIES. Three types of processes are typically used to accomplish the high-rate and stabilization phases: the windrow process in which composting occurs in long, parallel 1.8- to 2.4-m (6- to 8-ft) high and 3.7- to 4.9-m (12- to 16-ft) wide piles (windrows) that are constructed from the starting compost mix; the aerated static pile process that is similar to the windrow process but uses a concrete or asphalt pad as the reaction area; and in-vessel or reactor processes in which a constructed reactor that encases the mixture during the reaction period is used. See *Biosolids Composting* (WEF, 1995) for information about design and operation of each process.

MICROBIAL ACTIVITIES ASSOCIATED WITH COMPOSTING

The organisms primarily associated with decomposition during composting are *bacteria*. Mesophilic bacteria are those that grow optimally within the temperature range of 20 to 45 °C, whereas thermophilic bacteria grow optimally within the range of 45 to 80 °C and higher. Mesophiles initiate the composting process by metabolizing easily degradable compounds such as sugars, *amino acids*, carbohydrates, proteins, and lipids. As these nutrients are oxidized by bacteria, energy is released as heat, and the temperature rises. Although mesophiles do not grow at temperatures higher than approximately 45 °C, many are capable of metabolizing organic compounds at higher

Table 7.1 Representative organisms isolated from composting wastewater sludge with wood shavings as bulking agent.

Organism	Gram	High rate	Stabilization	Curing
Arthrobacter baumanii	+	+		+
Bacillus acidocaldarius	+		+	
Bacillus amyloliquifaciens	+		+	
Bacillus azotoformans	+	+		
Bacillus brevis	+	+	+	+
Bacillus coagulans	+	+	+	+
Bacillus KLC	+	+	+	+
Bacillus licheniformis	+	+	+	+
Bacillus megaterium	+	+	+	+
Bacillus pumilis	+	+	+	+
Bacillus sphaericus	+			+
Bacillus stearothermophilus	+	+	+	+
Bacillus subtilis	+	+	+	+
Bacillus thuringiensis	+		+	+
Comamonas testosteroni	–	+		
Enterococcus faecalis	+	+		
Gluconobacter cerinus	–	+		+
Klebsiella terrigena	–			+
Pseudomonas alcaligenes	–			+
Pseudomonas azeliaca	–			+
Pseudomonas fulva	–			+
Pseudomonas stutzeri	–	+		+
Pseudomonas syringae	–	+		
Rhodococcus erythropolis	+	+		
Rhodococcus globerulus	+			+
Xanthomonas campestris	–	+	+	

temperatures, and many thermophiles that do not grow at temperatures lower than 45 °C can degrade organic compounds in the mesophilic temperature range (Cook, 1993).

Both mesophiles and thermophiles can be isolated from compost at different times and during different phases in the composting cycle (Nagasaki et al., 1985a). Organisms capable of degrading various *macromolecules* such as protein, carbohydrates, cellulose, and constituents of bacterial cell walls are present throughout the composting process (Atkinson, 1995, and Atkinson et al., 1996). If the temperature exceeds 60 °C, the *species* diversity of the microbial community begins to decrease, and the efficiency of composting decreases (Strom, 1985).

There is no clear-cut microbial *succession* during composting but a constant flux of organisms (Table 7.1). Some organisms are found in high numbers throughout composting, some are detectable only when the temperature is in the mesophilic range, and some are found only toward the end of composting. Methods currently used to enumerate organisms are limited to detection of only the predominant organisms in the *microbial community*.

The levels of organisms isolated from compost are generally lower than would be predicted from the amount of carbon dioxide produced. This may be accounted for by cell turnover (some cells dying and being replaced by other growing cells) or by uncoupling of oxidative reactions from energy production in the cell (Nagasaki et al., 1985b). After composting, mesophiles and thermophiles remain viable for several months at levels of 10^7 to 10^8 colony forming units (CFU)/g of dry solids, both in dry and moist compost (Cook, 1993).

Filamentous *actinomycetes* have frequently been isolated from compost. They generally grow more slowly than other bacteria and degrade a variety of macromolecules such as *hemicellulose*, lipids, *starches*, and proteins and a variety of insoluble compounds such as cellulose, *chitin*, and *waxes*.

Fungi are also present in compost, where they are capable of degrading plant materials such as lignin and cellulose. They are usually found in greatest numbers in the outer layers of a compost pile, a likely result of their preference for lower temperatures and sufficient oxygen. Both fungi and actinomycetes are *obligate aerobes*. Fungi can be isolated from composting material but are usually present at lower levels than bacteria. They generally do not compete well with bacteria at temperatures associated with the high-rate phase (50 to 60 °C), but their levels increase during the curing phase, when cellulose and other plant polymers become the predominant substrate.

Protozoa, *rotifers*, and *algae* are typically destroyed during the thermophilic phase of composting and are not considered to be involved in composting. They may become reestablished during the curing phase.

CONDITIONS INFLUENCING MICROBIAL ACTIVITIES

The rate and extent of composting depends on maintaining physical and nutritional conditions conducive to optimal microbial activities (Composting Council, 1994, and Finstein and Hogan, 1993).

PARTICLE SIZE. Aerobic microbial activity is limited to the surfaces of particles. Therefore, small particle size (large surface-to-volume ratio) facilitates microbial degradation. If particles are too small (e.g., powdered sawdust), there will be insufficient free air space. The ideal compost mix represents a compromise between these criteria.

TEMPERATURE. The rate of decomposition of materials during composting increases as the temperature increases. However, when the temperature exceeds 55 to 60 °C, species diversity of metabolically active organisms decreases, and the decomposition process becomes less efficient (Finstein and

Hogan, 1993). The optimum temperature for composting is, therefore, between 55 and 60 °C. For composting of wastewater solids, it is desirable to maintain a high temperature for sufficient time to destroy pathogens (U.S. CFR).

AERATION. There are three reasons for providing air to the composting mixture (Finstein and Hogan, 1993). The first is to supply oxygen to the organisms that are decomposing the solids. Without oxygen, aerobic organisms cannot function, and subsequently *anaerobic* metabolic activities will occur, producing highly objectionable odors.

The second reason is to remove the excess heat generated by the organisms as a result of the biochemical reactions they are performing. If this heat is not removed, the temperature will increase until it reaches the point at which most organisms will no longer survive. One of the keys to operating a composting process successfully is to keep the temperature of the compost within the range at which the composting organisms function optimally. This is approximately between 55 and 60 °C. Control of the temperature of the compost mix is accomplished primarily through regulation of the quantity of airflow through the compost pile. Generally, much more air is required to remove heat than to provide oxygen. Therefore, temperature should be the controlling parameter to determine whether sufficient aeration is being provided.

The third reason is to remove excess moisture. The heat generated in the compost pile causes water in the mixture to evaporate. The resulting water vapor is removed by the air flowing through the compost pile. Water removal results in a product that is more easily handled and dry enough to be used as a bulking agent, if required. It may be necessary to add water if the material being composted becomes too dry. This often occurs in the area of air intake with in-vessel systems. When solids particles are too large, drying occurs on the outside of the particle, and the rate of metabolic activity is reduced. Remixing of the compost mix causes shearing of particles and *exposure* of new surfaces. Ideally, there should be separate aeration zones within each pile or reactor, and the air supply to each zone should be individually controlled. The direction of airflow should be reversible so that air can be pushed or pulled through the compost mix. Air supply can be controlled either by a timer or by signals from a temperature probe in the pile.

MOISTURE. Organisms require moisture for metabolic activity and are typically inactive if the moisture content of their environment is below 40%. Excessive moisture (>60%) is detrimental in composting because too much water reduces the free air space and results in oxygen limitation, leading to an odor problem. Therefore, maintaining a moisture level of 50 to 60% is important for successful composting.

NUTRIENTS AND CARBON-TO-NITROGEN RATIO. Organisms require carbon, nitrogen, sulfur, phosphorus, and a number of *trace* elements (e.g., iron, magnesium, and calcium) to grow. The amount of degradable organic carbon compounds determines the heat generated, which, in turn, determines the potential for pathogen destruction and the cost requirements for aeration. Only a portion of wastewater solids are degradable, and this fraction will depend on whether the solids are raw, digested aerobically, or digested anaerobically. It is important to remember that the bulking agent may also be degradable.

Nitrogen is the next nutrient that must be considered. A carbon-to-nitrogen ratio of approximately 30:1 is generally considered desirable in the starting mix. This provides enough nitrogen to support microbial growth (Haug, 1993). Ammonia-nitrogen can be lost by air stripping during composting. If the organic carbon content of the final compost is high relative to the available nitrogen, organisms consume the nitrogen, and it is no longer available to support plant growth. *Nitrification* has not been shown to occur at thermophilic temperatures.

The other required nutrients, including sulfur, phosphate, and trace minerals, are typically available in the wastewater solids. Phosphate might become limiting, depending on the method of phosphate removal during the wastewater treatment process.

pH. The optimum *pH* range for microbial activities is 6.5 to 8.5. Composting will occur if the pH of the starting material is outside this range (e.g., a pH of 5.5 to 11.0) but at a slower rate. Insufficient aeration can cause the pH to drop because of the production of organic acids. Increased aeration facilitates breakdown of the organic acids and production of carbon dioxide, which raises the pH. The pH is controlled by microbial activities and typically does not have to be adjusted during the composting of wastewater solids.

MICROBIAL INDICATORS OF COMPOST STABILITY

BIODEGRADABILITY OF SUBSTRATE AND BULKING AGENT. *Biodegradability* of the compost mix is a key factor in the design and operation of a compost system. It determines the energy available to heat the mix and the time required to achieve stabilization (Haug, 1993). Degradability of both the substrate and the bulking agent are important, and both can vary over a broad range, depending on the specific type of solids being composted (primary, aerobically digested or anaerobically digested) and the bulking agent being used.

During composting, a portion of the *organic matter* being composted is oxidized to carbon dioxide and water to yield the energy needed for cell *biosynthesis*, and a portion is incorporated to new cell *biomass*. The newly formed cell biomass can serve as a substrate for other cell-degrading bacteria. However, at cell densities of 10^7 to 10^8 cells per gram, the microbial cells represent only approximately 0.001 to 0.01% of the volume of material and are, therefore, a scarce source of nutrients.

Some of the material being composted is converted to humic substances, which represent a complex, long-term storage material for soil organisms. The progression from readily degradable nutrients to slowly degradable substrates to humic substances serving as nutrients represents a gradual transition with no sharp boundaries between stable and unstable composts.

RESPIRATION. The rate of oxygen uptake or carbon dioxide production is a measure of the level of microbial activity and organic nutrients available and is, therefore, a useful method for determining the degree of stabilization. An oxygen uptake rate of 5% or less of the starting rate with levels of 0.12 and 0.042 mg O_2/g dry solids•h for 8- and 10-month old composts, respectively, have been reported (Haug, 1993).

ODORS. Cured compost normally has a pleasant, musty odor because of the presence of actinomycetes. Unstable compost develops an unpleasant odor during storage. Odor control is of key concern for successful composting (Finstein and Hogan, 1993, and Walker, 1993).

PATHOGEN DESTRUCTION. Biological wastewater treatment does not completely remove or inactivate pathogens (Bitton, 1994). Potentially pathogenic organisms present in wastewater solids include bacteria, *viruses*, protozoan *cysts* and oocytes, and *helminthes* (Table 7.2). Elevated temperature is one of the most effective means of destroying or reducing the levels of these pathogens.

Although many pathogens are destroyed by temperatures of 55 °C for 3 days, some fungi, including *Aspergillus fumigatus*, can survive the composting process. *Aspergillus fumigatus* is an opportunistic pathogen that causes allergies when inhaled during composting operations and can cause lung damage. This organism is found in sawdust and wood shavings used as bulking agents and increases in numbers during the stabilization and curing phases of composting. High levels of *Aspergillus* have been reported in the air

Table 7.2 Pathogens commonly found in wastewater solids. (From *Wastewater Microbiology* by G. Bitton. Copyright © 1994 Wiley–Liss, New York. Reprinted by permission of Wiley–Liss, Inc., a subsidiary of John Wiley & Sons, Inc.)

Bacteria	Viruses	Protozoa	Helminthes
Salmonella	*Enteroviruses*	*Giardia*	*Ascaris*
Shigella	Hepatitis A	*Entamoeba*	*Taenia*
Campylobacter	Polio virus	*Cryptosporidium*	*Toxocara*
Yersinia	*Coxsackie* viruses	*Balantidium*	
Leptospira	*Echoviruses*	*Naegleria*	
Vibrio	*Adenoviruses*		
Escherichia coli (pathogenic)	*Caliciviruses*		
	Rotaviruses		
Streptococcus			

around composting facilities (10^3 to 10^4 CFU/g). Care should be taken to minimize exposure to pathogens at composting sites (Composting Council, 1994), and both Occupational Safety and Health Administration and National Institute of Occupational Safety and Health regulations should be reviewed and followed.

REFERENCES

Alexander, M. (1977) *Introduction to Soil Microbiology.* Wiley & Sons, New York.

Atkinson, C.M.F. (1995) *Biodegradabilities and Microbial Activities During Composting of Solid Wastes.* Ph.D. dissertation, Univ. Alab. Birmingham, Birmingham, Ala.

Atkinson, C.M.F.; Jones, D.D.; and Gauthier, J.J. (1996) Biodegradabilities and Microbial Activities During Composting of Oxidation Ditch Sludge. *Compost Sci. Utiliz.*, **4**, 84.

Bitton, G. (1994) *Wastewater Microbiology.* Wiley–Liss, New York.

Composting Council (1994) *Compost Facility Operating Guide.* Alexandria, Va.

Cook, K.L. (1993) *Microbial Ecology of Composting Under Various Physical Conditions.* Masters thesis, Univ. Alab. Birmingham, Birmingham, Ala.

Finstein, M.S., and Hogan, J.A. (1993) Integration of Composting Process Microbiology, Facility Structure and Decision Making. In *Science and Engineering of Composting: Design, Environmental Microbiological and Utilization Aspects.* H.A.J. Hoitink and H.A. Keener (Eds.), Renaissance Publications, Worthington, Ohio, 1.

Haug, R.T. (1993) *The Practical Handbook of Compost Engineering.* Lewis Publishers, Boca Raton, Fla.

Nagasaki, K.; Sasaki, M.; Shoda, M.; and Kubota, H. (1985a) Characteristics of Mesophilic Bacteria Isolated During Thermophilic Composting of Sewage Sludge. *Appl. Environ. Microbiol.*, **49**, 42.

Nagasaki, M.; Shoda, M.; and Kubota, H. (1985b) Effect of Temperature on Composting of Sewage Sludge. *Appl. Environ. Microbiol.*, **50**, 1526.

Strom, P.F. (1985) Effect of Temperature on Bacterial Species Diversity in Thermophilic Solid-Waste Composting. *Appl. Environ. Microbiol.*, **50**, 899.

U.S. Code of Federal Regulations. *The Standards for the Use or Disposal of Sewage Sludge.* Title 40, 40 CFR, Part 503. 58 *Fed. Regist.* 9248.

Walker, J.M. (1993) Control of Composting Odors. In *Science and Engineering of Composting: Design, Environmental Microbiological and Utilization Aspects.* H.A.J. Hoitink and H.A. Keener (Eds.), Renaissance Publications, Worthington, Ohio, 185.

Water Environment Federation (1995) *Biosolids Composting.* Special Publication, Alexandria, Va.

SUGGESTED READINGS

American Public Health Association; American Water Works Association; and Water Environment Federation (1992) *Standard Methods for the Examination of Water and Wastewater.* 18th Ed., Washington, D.C.

Spohn, E. (1969) How Ripe Is Compost? *Compost Sci.*, **10**, 3.

Walker, J.; Knight, K.; and Stein, L. (1994) *A Plain English Guide to the EPA Part 503 Biosolids Rule.* EPA-832/R-93-003, U.S. EPA, Washington, D.C.

Williams, T.O., and Miller, F.C. (1993) Composting Facility Odor Control Using Biofilters. In *Science and Engineering of Composting: Design, Environmental Microbiological and Utilization Aspects.* H.A.J. Hoitink and H.A. Keener (Eds.), Renaissance Publications, Worthington, Ohio, 262.

Glossary

acclimation The dynamic response of a system to the addition or deletion of a substance, until equilibrium is reached; adjustment to a change in the environment.

acetic acid CH_3COOH, a clear, colorless liquid or crystalline mass with a pungent odor, miscible with water or alcohol; crystallizes in deliquescent needles; a component of vinegar. Also known as ethanoic acid.

acetogenesis Microbial conversion of carbon dioxide to sugars and acetate.

acidophyle (acidophile) An organism that grows best under acid conditions (down to a pH of 1).

actinomycete Highly filamentous bacteria; moldlike bacteria.

activated sludge Sludge withdrawn from a secondary clarifier following the activated sludge process; consists mostly of biomass with some inorganic settable solids. Return sludge is recycled to the head of the process; waste (excess) sludge is removed for conditioning.

active transport The pumping of ions or other substances across a cell membrane against an osmotic gradient, that is, from a lower to higher concentration.

acute Having a sudden onset, sharp rise, and short course.

adaptation The occurrence of genetic changes in a population or species as the result of natural selection so that it adjusts to new or altered environmental conditions.

adhesion Intimate sticking together of surfaces.

adsorption The adherence of a material to the surface of a solid.

adsorption coefficient A measure of a material's tendency to absorb to soil or other particles.

aeration The addition of oxygen to water or wastewater, usually by mechanical means, to increase dissolved oxygen levels and maintain aerobic conditions.

aerobe An organism that requires air or free oxygen to maintain its life processes.

aerobic Requiring, or not destroyed by, the presence of free elemental oxygen.

aerobic oxidation (respiration) The conversion of organics such as carbohydrates to bacterial cells, carbon dioxide, water, and energy.

aerosol A suspension of colloidal particles in air or another gas.

aerotolerant anaerobe Microbes that grow under both aerobic and anaerobic conditions but do not shift from one mode of metabolism to another as conditions change. They obtain energy exclusively by fermentation.

agar A gelatinous substance extracted from red algae; commonly used as a medium for laboratory cultivation of bacteria.

agglutinate To unite or combine into a group or mass.

aggregate Crowded or massed into a dense cluster.

alcohol Any of a class of organic compounds containing the hydroxyl group –OH.

aldehyde One of a class of organic compounds containing the CHO radical.

algae Photosynthetic microscopic plants, which, in excess, can contribute taste and odor to potable water and deplete dissolved oxygen on decomposition.

aliphatic Of or pertaining to any organic compound of hydrogen and carbon characterized by a straight chain of the carbon atoms; the subgroups of such compounds are alkanes, alkenes, and alkynes.

alkaline Having a pH greater than 7.

alkalinity The alkali concentration or alkaline quality of an alkali-containing substance.

alkane A member of a series of saturated aliphatic hydrocarbons having the empirical formula C_nH_{2n+2}.

alkene One of a class of unsaturated aliphatic hydrocarbons containing one or more carbon-to-carbon double bonds.

alkyne One of a group of organic compounds containing a carbon-to-carbon triple bond.

aloricate Without a lorica (a hard protective shell or case).

ambient The environmental conditions in a given area.

amination A process in which the amino group (–NH_2) is introduced to organic molecules.

amino acid Any of the organic compounds that contain one or more basic amino groups and one or more acidic carboxyl groups and that are polymerized to form peptides and proteins; only 20 of the more than 80 amino acids found in nature serve as building blocks for proteins; examples are tyrosine and lysine.

amoeba Protozoa that uses pseudopodia (temporary extensions of the cell) for locomotion and feeding.

amorphous Pertaining to a solid that is noncrystalline, having neither definite form nor structure.

anabiosis State of suspended animation induced by desiccation and reversed by addition of moisture.

anabolic Pertaining to the part of metabolism involving the union of smaller molecules into larger molecules; the method of synthesis of tissue structure.

anabolism A part of metabolism involving the union of smaller molecules into larger molecules; the method of synthesis of tissue structure.

anaerobic (1) A condition in which no free oxygen is available. (2) Requiring, or not destroyed by, the absence of air or free oxygen.

anaerobic respiration Respiration under anaerobic conditions. The terminal electron acceptor, instead of oxygen in the case of regular respiration, can be CO_2, Fe^{2+}, fumarate, nitrate, nitrite, nitrous oxide, sulfur, sulfate, etc. Note that anaerobic respiration still uses an electron transport chain to dump the electron while fermentation does not.

anaerobiosis A mode of life carried on in the absence of molecular oxygen.

anatomical Relating to the structure of organisms.

anion An ion that is negatively charged.

annelid A multisegmented worm.

anoxic Lacking in oxygen.

anoxygenic See anaerobic.

antagonistic Mutual opposition; nullification by one substance of a chemical or physical effect due to another.

anterior Situated near or toward the front or head of an animal body.

anthrone test A test to determine total carbohydrates; the reagent hydrolyzes polysaccharides to monosaccharides and forms a colored compound in the presence of monosaccharides.

anthropogenic Referring to environmental alterations resulting from the presence or activities of humans.

antibiotic A substance produced by a microorganism, which, in dilute solution, inhibits or kills another microorganism.

antibody Any of the body of globulins (proteins) that combine specifically with antigens and neutralize toxins.

anus The posterior orifice of the alimentary canal.

aperture An opening, for example, the opening in a photographic lens that admits light.

aquaculture Cultivation of natural faunal resources of water.

aqueous Relating to or made with water.

aromatic Pertaining to or characterized by the presence of at least one benzene ring.

artificial support media Material such as a glass slide or polyurethane foam that is introduced to the wastewater treatment system to allow colonization by protozoa for later collection and examination.

asexual reproduction Reproduction (cell division, spore formation, fission, or budding) without union of individuals or germ cells.

assimilation Conversion of nutritive materials to protoplasm.

asthma A pulmonary disease marked by labored breathing, wheezing, and coughing; cause may be emotional stress, chemical irritation, or exposure to an allergen.

Aufwuchs community A community of microscopic plants and animals associated with the surfaces of submerged objects.

autoclavable Able to be sterilized by autoclaving.

autolysin An enzyme that causes the cell that made it to self-destruct.

autotroph An organism capable of synthesizing organic nutrients directly from simple inorganic substances, such as carbon dioxide and inorganic nitrogen.

bacilloid Rod shaped.

bacteria A group of universally distributed, rigid, essentially unicellular microscopic organisms lacking chlorophyll. They perform a variety of biological treatment processes, including biological oxidation, sludge digestion, nitrification, and denitrification.

bacterivorous Feeding on bacteria.

baseline A sample used as comparative reference point when conducting further tests or calculations.

basophyle (basophile) An organism that prefers, or can tolerate, alkaline conditions, typically in the pH range 8 to 11.

bench-scale testing A small-scale test or study used to determine whether a technology is suitable for a particular application.

benthic Relating to the bottom or bottom environment of a body of water.

beta (ß) oxidation The process of fatty-acid catabolism in which two-carbon fragments are removed in succession from the carboxyl end of the chain.

binary fission Form of asexual reproduction in some microbes in which the parent organism splits into two independent organisms.

binocular eyepiece Device that divides a beam of light into two different beams and directs them through two eyepiece tubes so that both eyes can be used at the microscope.

bioactivation Transformation of a chemical within an organism to a biochemically active metabolite.

bioaugmentation The addition of commercially prepared cultures of saprophytic or nitrifying bacteria.

bioavailability The degree to which an agent, such as a drug or nutrient, becomes available at the physiological site of activity.

biochemical oxygen demand (BOD) A standard measure of wastewater strength that quantifies the oxygen consumed in a stated period of time, usually 5 days, and at 20 °C.

biodegradability The characteristic of a substance to be broken down by organisms.

biodegradation The breaking down of a substance by organisms.

bioenergetics The branch of biology dealing with energy transformations in living organisms.

biofilm An accumulation of microbial growth.

bioindicator Species or group of species that is representative and typical for a specific status of an ecosystem, that appears frequently enough to serve for monitoring, and whose population shows a sensitive response to changes.

biological transformation A chemical transformation within a living cell.

biomass The dry weight of living matter, including stored food, present in a species population and expressed in terms of a given area or volume of the habitat.

biomass carriers An inert material on which organisms can grow while feeding on wastewater organics and other materials.

biooxidation The process by which living organisms, in the presence of oxygen, convert organic matter to a more stable or a mineral form. *(The process by which all living things obtain energy for metabolic processes.)*

biorecalcitrant Resistant to biological treatment.

biosynthesis Production, by synthesis or degradation, of a chemical compound by a living organism.

bivalve shell Having a shell composed of two valves.

blank sample Laboratory simulated test material known to be free of the chemical being analyzed. A portion of the blank material is used to test the method, apparatus, and reagents for interferences or contamination.

bog Soft, waterlogged ground; marsh.

bottomland Low-lying land along a river.

brackish (1) Of water, having salinity values ranging from approximately 0.50 to 17.00 parts per thousand. (2) Of water, having less salt than sea water, but undrinkable.

bristleworm Aquatic annelid of the class Polychaeta that has both errant and sedentary species.

Brownian motion Random movements of small particles suspended in a fluid, caused by the statistical pressure fluctuations over the particle.

buccal opening Cavity of the mouth.

buffering capacity The relative ability of a buffer solution to resist pH change with addition of an acid or base.

bulking Clouds of billowing sludge that occur throughout secondary clarifiers and sludge thickeners when the sludge does not settle properly. In the activated sludge process, bulking is usually caused by filamentous bacteria or bound water.

bulking sludge Low-density activated sludge that settles poorly.

butyric acid $CH_3CH_2CH_2COOH$, a colorless, combustible liquid with a boiling point of 163.5 °C; soluble in water, alcohol, and ether; used in the synthesis of flavors, in pharmaceuticals, and in emulsifying agents.

calcareous Containing calcium.

carapace A bony or chitinous case or shield covering the back or part of the back of an animal.

carbohydrates Any of the group of organic compounds composed of carbon, hydrogen, and oxygen, including sugars, starches, and celluloses.

carbon fixation A process occurring in photosynthesis whereby atmospheric carbon dioxide gas is combined with hydrogen obtained from water molecules.

carbonaceous biochemical oxygen demand The portion of biochemical oxygen demand whereby oxygen consumption is caused by oxidation of carbon; usually measured after a sample has been incubated for 5 days.

carboxylic acid Any of a family of organic acids characterized by the presence of one or more carboxyl groups.

carcinogen Any agent that incites development of a carcinoma or any other sort of malignancy.

carrier protein A protein that transports specific substances through the cell membrane in which it is embedded and to the cell. Different carrier proteins are required to transport different substances because each one is designed to recognize only one substance or group of similar substances.

catabolic Pertaining to the metabolic change of complex to simple molecules.

catabolism That part of metabolism concerned with the breakdown of large protoplasmic molecules and tissues, often with the liberation of energy.

catalyze To modify the rate of a chemical reaction as a catalyst.

cation An ion that is positively charged.

caudal Toward, belonging to, or pertaining to the tail or posterior end.

causative Effective or operating as a cause or agent.

caustic Alkaline or basic.

cavitation Pitting of a solid surface such as metal or concrete.

cellulolytic Of, pertaining to, or causing the hydrolysis of cellulose.

cellulose The main polysaccharide in living plants, forming the skeletal structure of the plant cell wall; a polymer of ß-D-glucose units linked together, with the elimination of water for form chains composed of 2000 to 4000 units.

centrate The liquid remaining after solids have been removed in a centrifuge.

cephalothorax The body division comprising the united head and thorax of arachnids and higher crustaceans.

chain-of-custody Documentation of times, dates, and personnel involved in sample collection, transport, and analysis.

chelating agent An organic compound in which atoms form more than one coordinate bond with metals in solution.

chelation A chemical process involving formation of a heterocyclic ring compound that contains at least one metal cation or hydrogen ion in the ring.

chemisorption A chemical adsorption process in which weak chemical bonds are formed between gas or liquid molecules and a solid surface.

chemoautotroph Any of a number of autotrophic bacteria and protozoans that do not carry out photosynthesis.

chemoheterotroph An organism that derives energy and carbon from the oxidation of preformed organic compounds.

chemolithotroph An organism that obtains its energy from the oxidation of inorganic compounds.

chemoorganotroph An organism that requires an organic source of carbon and metabolic energy.

chemotroph An organism that extracts energy from organic and inorganic oxidation–reduction reactions.

chitin A white or colorless amorphous polysaccharide that forms a base for the hard outer integuments of crustaceans, insects, and other invertebrates.

chlorination The addition of chlorine to water or wastewater, usually for the purpose of disinfection.

chlorine Cl_2, an oxidant commonly used as a disinfectant in water and wastewater treatment.

chlorophyll The generic name for one of several plant pigments that function as photoreceptors of light energy for photosynthesis.

chloroplast A plastid that contains chlorophyll and is the site of photosynthesis.

cholesterol A sterol produced by all vertebrate cells, particularly in the liver, skin, and intestine, and found most abundantly in nerve tissue.

chromatography A method of separating and analyzing mixtures of chemical substances by chromatographic adsorption.
elution—the removal of absorbed species from a porous bed or chromatographic column by means of a stream of liquid or gas.
gas—a separation technique involving passage of a gaseous moving phase through a column containing a fixed absorbent phase; used principally as a quantitative analytical technique for volatile compounds.
high performance liquid—a laboratory technique, a type of column chromatography that uses a combination of several separation techniques to separate substances at a high resolution.
thin-layer—chromatography using a thin layer of powdered medium on an inert sheet to support the stationary phase.

chronic Long continued; of long duration.

cilia Relatively short centriole-based, hairlike processes on certain anatomical cells and motile organisms.

ciliate Any of various protozoans of the phylum Ciliophora having cilia at some stage in their life cycle.

clarifier A quiescent tank used to remove suspended solids by gravity settling. Also called sedimentation or settling basins, they are usually equipped with a motor-driven chain-and-flight or rake mechanism to collect settled sludge and move it to a final removal point.

cleavage The state of being split or cleft; a fissure or division.

cloaca The chamber that functions as a respiratory, excretory, and reproductive duct in certain invertebrates.

coagulant Chemical added to destabilize, aggregate, and bind together colloids and emulsions to improve settleability, filterability, or drainability.

coccoid (cocci) A spherical bacterial cell.

coenzyme The nonprotein portion of an enzyme; a prosthetic group that functions as an acceptor of electrons or functional groups.

coliform bacteria Rod-shaped bacteria living in the intestines of humans and other warm-blooded animals.

colloid Suspended solid with a diameter smaller than 1 m that cannot be removed by sedimentation alone.

colloidal material Finely divided solids that will not settle but may be removed by coagulation, biochemical action, or membrane filtration.

colonization The establishment of immigrant species in a peripherally unsuitable ecological area; occasional gene exchange with the parental population occurs, but generally the colony evolves in relative isolation and in time may form a distinct unit.

co-metabolism The metabolic transformation of a substance while a second substance serves as primary energy or carbon source.

commensal An organism living in a state of *commensalism (an interspecific, symbiotic relationship in which two different species are associated, wherein one is benefitted and the other is neither benefitted nor harmed).*

community Aggregation of organisms characterized by a distinctive combination of two or more ecologically related species; an example is a deciduous forest.

compactability A measure of the ability of a substance to be compacted, which increases bulk density and decreases porosity.

competition The inter- or intraspecific interaction resulting when several individuals share an environmental necessity.

competitive exclusion The result of a competition in which one species is forced out of part of the available habitat by a more efficient species.

competitive exclusion principle (Gause's principle) A statement that two species cannot occupy the same niche simultaneously.

complexation See complexing.

complexing Formation of a complex compound.

composting A managed method for the biological decomposition of organic materials frequently used to stabilize sludge.

compound microscope Microscope consisting of an objective and eyepiece mounted in a drawtube.

concentrator A plant in which materials are concentrated.

condenser A lens or mirror used to shorten or diminish in size.

confluence (1) A stream formed from the flowing or two or more streams. (2) The place where such streams form.

confocal microscopy A system of (usually) epifluorescent light microscopy in which a fine laser beam of light is scanned over the object through the objective lens. The technique is particularly good at rejecting light from outside the plane of focus and thus produces higher effective resolution than is normally achieved.

conjugation A process involving contact between two bacterial or ciliate cells during which genetic material is passed from one cell to the other.

conserve To protect from loss or depletion.

constitutive Making a thing what it is; essential.

contractile Having the power to shorten or diminish in size.

contractile vacuole A tiny, intracellular membranous bladder that functions in maintaining intra- and extracellular osmotic pressures in equilibrium as well as excretion of water, such as occurs in protozoans.

coprozoic Living in fecal matter.

copulation The sexual union of two individual organisms, resulting in insemination or deposition of the male gametes in close proximity to the female gametes.

corona The upper portion of a body part (as a tooth or the skull).

corrosion Gradual destruction of a metal or alloy due to chemical processes such as oxidation or the action of a chemical agent.

coryneform (1) Of bacteria, rod shaped with one end substantially thicker than the other. (2) Any Gram-positive, nonmotile bacteria that occur as irregularly shaped rods and resemble members of the genus *Corynebacterium*.

covalent bond A bond in which each atom of a bond pair contributes one electron to form a pair of electrons.

crustaceans Any of a large class of mostly aquatic mandibulate arthropods that have a chitinous or calcerous and chitinous exoskeleton, a pair of often modified appendages on each segment, and two pairs of antennae and that include lobsters, shrimps, crabs, wood lice, water fleas, and barnacles.

cryptobiosis (cryptobiotic state) A state in which the metabolic rate of an organism is reduced to an imperceptible level.

cuticle An external layer usually secreted by epidermal cells.

cyanobacteria Blue-green algae.

cycle Process in which the system returns to the original point at the end of a complete operation, for example, nitrogen cycle in plant life.

cyrtophorid An ostracod crustacean larva with a long first pair of antennae and lost swimming capability in the second pair. Also, a group of ciliate protozoa.

cyst Resting stage formed by some bacteria, nematodes, and protozoa in which the whole cell is surrounded by a protective layer, not the same as endospore.

cytomembrane The internal membrane found in nitrifying bacteria in biofilm.

cytopharyngeal A region of the plasma membrane or cytoplasm of some ciliated and flagellated protists specialized for endocytosis; a permanent oral canal.

deanimate Removal from a molecule of the amino group.

decarboxylate To remove the carboxyl radical, especially from amino acids and protein.

dechlorinating agent A chemical added to a solution to neutralize or bind excess chloride.

definitive host The host in which a parasite reproduces sexually.

degradation The breakdown of substances by biological action.

deleterious Having a harmful effect, injurious.

denitrification The reduction of nitrate or nitrite to gaseous products such as nitrogen, nitrous oxide, and nitric oxide; brought about by denitrifying bacteria.

depolymerize Decomposition of macromolecular compounds into relatively simple compounds.

derivatization The conversion of a chemical compound to a derivative.

desiccation Thorough removal of water from a substance, often with the use of a desiccant.

detention time The theoretical time required to displace the contents of a tank or unit at a given rate of discharge.

detritus Disintegrated matter.

diabetes Any of various abnormal conditions characterized by excessive urinary output, thirst, and hunger.

diaphragm Any opening in an optical system that controls the cross section of a beam of light passing through it to control light intensity, reduce aberration, or increase depth of focus. Also known as lens stop.

diatom Unicellular, microscopic algae with a boxlike structure consisting principally of silica.

dichotomous keys Scheme for making identifications by asking a series of questions that can be answered directly and without quantification, for example, motile, rod shaped, spore produced?

didymium A mixture of rare-earth elements made up chiefly of neodymium and praseodymium and used especially for coloring glass for optical fibers.

diffraction A modification light undergoes in passing by the edges of opaque bodies or through narrow slits or in being reflected from ruled surfaces and in which the rays appear to be deflected and produce fringes of parallel light and dark-colored bands.

diffusion The spatial equalization of one material throughout another.

diffusion coefficient The weight of a material in grams diffusing across an area of 1 cm^2 in 1 second in a unit concentration gradient. Also known as diffusivity.

diffusivity See diffusion coefficient.

digester A tank or vessel used for sludge digestion.

digestion The process of converting food to an absorbable form by breaking it down to simpler chemical compounds.

dilute To make thinner or less concentrated by adding a liquid such as water; to lessen the force, strength, purity, or brilliance of, especially by admixture.

dilution rate The reciprocal of hydraulic retention time.

dinoflagellate Any of a class (Dinoflagellatea) of unicellular flagellate protozoa that include luminescent forms, forms important in marine food chains, and forms causing red tide.

dipeptide A peptide that yields two molecules of amino acid on hydrolysis.

diphtheria An acute febrile contagious disease marked by the formation of a false membrane especially in the throat and caused by a bacterium (*Corynebacterium diphtheriae*) that produces a toxin causing inflammation of the heart and nervous system.

disaccharide Any of the class of compound sugars that yield two monosaccharide units on hydrolysis.

discoid Being flat and circular in form.

disinfection With a disinfectant, the killing or reduction in numbers of waterborne fecal and pathogenic bacteria and viruses in potable water supplies or wastewater effluents.

dispersion Scattering and mixing.

dissimilation A form of microbial metabolism, oxidative in nature; breakdown.

dissimilative sulfate reduction Generation of sulfate from hydrogen sulfide.

dissolved oxygen The oxygen dissolved in a liquid.

distilled water Water that has been freed of dissolved or suspended solids and organisms by distillation.

dominance The influence that a controlling organism has on numerical composition or internal energy dynamics in a community.

dry mass The mass of organisms calculated after removal from the media and drying of the culture.

dysentery Inflammation of the intestine characterized by pain, intense diarrhea, and the passage of mucus and blood.

ecosystem A functional system that includes the organisms of a natural community together with their environment.

ectoplasm The outer relatively rigid granule-free layer of the cytoplasm usually held to be a gel reversibly convertible to a sol.

effluent Partially or completely treated water or wastewater flowing out of a basin or treatment plant.

electrolyte A chemical compound that, when molten or dissolved in certain solvents (usually water), will conduct an electric current.

electron acceptor An atom or part of a molecule joined by a covalent bond to an electron donor.

electron donor An atom or part of a molecule that supplies both electrons of a duplet forming a covalent bond.

electron microscope A microscope in which a beam of electrons focused by means of electromagnetic lenses is used to produce an enlarged image of a minute object on a fluorescent screen or photographic plate.

electron transport system The components of the final sequence of reactions in biological oxidations; composed of a series of oxidizing agents arranged in order of increasing strength and terminating in oxygen.

electrophile An electron-deficient ion or molecule that takes part in an electrophilic process.

electrostatic properties Pertaining to electricity at rest, such as an electric charge on an object.

elimination To get rid of, remove; to excrete as waste.

eluant A liquid used to extract one material from another, as in chromatography.

elute To extract one material from another, usually by means of a solvent.

embryonic state A state resembling or having characteristics of an embryo.

emulsion A stable dispersion of one liquid in a second immiscible liquid.

encephalitis Inflammation of the brain.

encyst The process of forming or becoming enclosed in a cyst or capsule.

endemic Restricted or native to a particular locality or region.

endocarditis Inflammation of the endocardium.

endoplasm The inner relatively fluid part of the cytoplasm (the protoplasm) of a cell external to the nuclear membrane.

endospore Differentiated cell formed within the cells of certain Gram-positive bacteria and extremely resistant to heat and other harmful agents.

enrichment A process that changes the isotopic ratio in a material.

enteric Of or within the intestine.

enterococci A group of sphere-shaped bacteria that normally inhibit the intestines of humans or other animals.

enteropathogenic Tending to produce disease in the intestinal tract.

enumeration To determine the number of, count.

enzyme Any of a group of catalytic proteins that is produced by living cells and that mediates and promotes the chemical processes of life without itself being altered or destroyed.

epidemic A disease that occurs simultaneously in a large fraction of the community.

epidemiological Pertaining to the study of the incidence, distribution, and control of disease in a population.

esophagus The tubular portion of the alimentary canal between the pharynx and the stomach.

ester The compound formed by the elimination of water and the bonding of an alcohol and an organic acid.

estuarine Of, relating to, or formed in an estuary.

eubacteria A superclassification (above kingdom level) of all prokaryotes, excludes Archaebacteria.

euglenoid Any member of the division Euglenophyta.

eukaryote A cell with a definitive nucleus.

eutrophication Nutrient enrichment of a lake or other water body, typically characterized by increased growth of planktonic algae and rooted plants. It can be accelerated by wastewater discharges and polluted runoff.

execretory system Those organs concerned with solid, fluid, or gaseous excretion.

exoenzyme An enzyme that functions outside the cell in which it is synthesized.

exogeneous Due to an external cause; not arising within the organism.

exoskeleton The external supportive covering of certain invertebrates, such as arthropods.

exotoxin A soluble poisonous substance given off during the growth of an organism.

exposure (1) The act or fact of exposing or being exposed. (2) The product of the duration and intensity of light striking a photosensitive material; the former is controlled by shutter speed, the latter by aperture.

extracellular organic binding substance Insoluble polysaccharides.

extracellular polymers Insoluble "slime" of polysaccharides.

extreme thermophile (hyperthermophile) An organism living at temperatures near 110 °C.

eyespot A simple light-sensitive organ of pigment or pigmented cells covering a sensory termination.

facilitated diffusion The movement of polar molecules across a cell membrane via protein transporters.

facultative anaerobe A microorganism that grows equally well under aerobic and anaerobic conditions.

fauna Animals.

fermentation Changes in organic matter or organic wastes brought about by anaerobic microorganisms and leading to the formation of carbon dioxide, organic acids, or other simple products.

fibril A small filament or fiber.

field-of-view eyepiece A special eyepiece equipped with an indicator for the actual area being photographed.

filamentous microorganisms Bacteria and fungi that grow in threadlike colonies resulting in a biological mass that will not settle and may interfere with drainage through a filter.

filiform Threadlike or filamentous.

flagella Relatively long, whiplike, centriole-based locomotor organelles on some motile cells.

flagellate An organism that propels itself by means of a relatively long, whiplike, centriole-based locomotor organelle; a flagella.

flatworm The common name for members of the phylum Platyhelminthes; individuals are dorsoventrally flattened.'

floc Collections of smaller particles agglomerated into larger, more easily settleable particles through chemical, physical, or biological treatment.

flocculation In water and wastewater treatment, the agglomeration of colloidal and finely divided suspended matter after coagulation by gentle stirring by either mechanical or hydraulic means. In biological wastewater treatment when coagulation is not used, agglomeration may be accomplished biologically.

fluorescent Exhibiting or capable of the emission of electromagnetic radiation, especially light.

food chain The scheme of feeding relationships by trophic levels, which unites the member species of a biological community.

food-to-microorganism ratio (F:M) In the activated sludge process, the loading rate expressed as the amount of biochemical oxygen demand per amount of mixed liquor suspended solids per day.

food vacuole A membrane-bound organelle in which digestion occurs in cells capable of phagocytosis.

food web A modified food chain that expresses feeding relationships at various, changing trophic levels.

formic acid HCOOH, a colorless, pungent, toxic, corrosive liquid melting at 8.4 °C; soluble in water, ether, and alcohol; used as a chemical intermediate and solvent in dyeing and electroplating processes and in fumigants. Also known as methanoic acid.

free-living Living or moving independently.

free radical An atom or a diatomic or polyatomic molecule that possesses at least one unpaired electron.

fretting Surface damage usually in an air environment between two surfaces, one or both of which are metals in close contact under pressure and subject to a slight relative motion.

f-stop An aperture setting for a camera lens; indicated by the f-number.

fulvic acids A group of organic acids formed in water on degradation of organic plant matter.

fungi Nucleated, usually filamentous, spore-bearing organisms devoid of chlorophyll.

gastroenteritis An inflammation of the stomach and intestinal tract.

gastrotrich Any of a phylum (Gastrotricha) of minute aquatic pseudocoelomate animals that usually have a spiny or scaly cuticle and cilia on the ventral surface.

gelatin A protein derived from the skin, white connective tissue, and bones of animals.

gene The basic unit of inheritance.

genus (plural genera) A taxonomic category that includes groups of closely related species; the principal subdivision of a family.

geochemistry The study of the chemical composition of the various phases of the earth and the physical and chemical processes that have produced the observed distribution of elements and nuclides in these phases.

geosmin A metabolite of certain blue-green algae and actinomycetes that causes an earthy–musty taste and odor in water supplies.

glucose $C_6H_{12}O_6$, a monosaccharide; occurs free or combined and is the most common sugar.

glycocalyx The outer component of a cell surface, outside the plasmalemma; usually contains strongly acidic sugars, hence it carries a negative electric charge.

glycolysis The enzymatic breakdown of glucose or other carbohydrate, with the formation of lactic acid or pyruvic acid and the release of energy in the form of adenosinetriphosphate.

glycoprotein A conjugated protein in which the nonprotein group is a carbohydrate.

gonidia A sexual reproductive cell or groups of cells arising in a special organ on or in a gametophyte.

graduated cylinder A tall narrow container with a volume scale used especially for measuring liquids.

graininess Mottled (spotted) appearance of an exposed photograph.

Gram-negative bacteria Bacteria that do not retain the purple dye used in the Gram staining method due to a higher lipid content of the cell wall.

Gram-positive bacteria Bacteria that retain the purple dye used in the Gram staining method.

Gram stain A common staining procedure used to differentiate bacteria to Gram-negative and Gram-positive categories.

granule A small grain or pellet; particle.

graticule A scale on transparent material in the focal plane of an optical instrument for locating and measuring objects.

gravity separation A process in which the components of a nonhomogeneous mixture separate themselves by the force of gravity in accordance with their individual densities.

group translocation A process of actively importing compounds to the bacterial cell. The compound diffuses into the cell passively and is immediately modified (for example, by phosphorylation) so that it cannot diffuse back out.

growth factors Any factor, genetic or extrinsic, that affects growth.

growth rate Increase in the number of bacteria in a population per unit time.

gymnostome Any of a class of ciliates (Gymnostomatea) characterized by having simple or inconspicuous oral cilia and usually with extrusomes for food capture.

habitat The part of the physical environment in which a plant or animal lives.

halobacteria Bacteria that live in conditions of high salinity.

halogen Any of the elements of the halogen family consisting of fluorine, chlorine, bromine, iodine, and astatine.

heavy metal A metal that can be precipitated by hydrogen sulfide in an acid solution and that may be toxic to humans in excess of certain concentrations.

helminth Any parasitic worm.

hemacytometer A specifically designed, ruled, and calibrated glass slide used with a microscope to count organisms.

hemicellulose A type of polysaccharide found in plant cell walls in association with cellulose and lignin; soluble in and extractable by dilute alkaline solutions.

Henry's law The mass of a gas dissolved by a given volume of liquid at a constant temperature is proportional to the pressure of the gas.

hepatitis An acute viral disease that results in liver inflammation and may be transmitted by direct contamination of the water supply by wastewater.

hermaphroditic Having the sex organs and many of the secondary sex characteristics of both male and female.

heterogeneous Consisting of or involving dissimilar elements or parts; not homogeneous.

heterotroph An organism that obtains nourishment from the ingestion and breakdown of organic matter.

heterotrophic bacteria A type of bacteria that derives its cell carbon from organic carbon; most pathogenic bacteria are heterotrophic bacteria.

homogeneous Pertaining to a substance having uniform composition or structure.

homogenize To make homogeneous.

hopper A funnel-shaped receptacle with an opening at the top for loading and a discharge opening at the bottom for bulk delivery of material.

humic substances (humus) Substances pertaining to or derived from humus.

humus Total of the organic compounds in soil exclusive of undecayed plant and animal tissues, their "partial decomposition" products, and the soil biomass. The term is often used synonymously with soil organic matter.

hydraulic retention time (HRT) (hydraulic residence time) The average time of retention of liquid in the treatment system. Vessel volume divided by the rate of liquid removed, resulting in a unit of time.

hydraulic shear Force exerted by water flowing past a surface that works to remove objects adhered to that surface.

hydric Pertaining to, characterized by, or requiring considerable moisture.

hydride A compound containing hydrogen and another element.

hydrocarbon An organic compound consisting predominantly of carbon and hydrogen.

hydrogen peroxide H_2O_2, unstable, colorless, heavy liquid boiling at 158 °C; soluble in water and alcohol; used as a bleach, chemical intermediate, rocket fuel, and antiseptic.

hydrology The science that treats the occurrence, circulation, distribution, and properties of the waters of the earth and their reactions with the environment.

hydrolysis (1) The reaction of a solute with water in aqueous solution. (2) A change in the chemical composition of matter produced by combination with water. Sometimes loosely applied in wastewater practice to the liquefaction of solid matter in a tank as a result of biochemical activity. (3) Usually a chemical degradation of organic matter; also the combination of water with metallic cations; also used in reference to the amount of anionic groups in certain flocculants.

hydrolyze To split complex molecules into simpler molecules by chemical reaction with water.

hydrophilic Having an affinity for, attracting, adsorbing, or absorbing water.

hydrophobic Lacking an affinity for, repelling, or failing to adsorb or absorb water.

hydrophytic Plants that require large amounts of water for growth.

hygienic practices Practices tending to promote or preserve health.

hymenostome Any of a subclass of ciliates in the class Oligohymenophorea having fairly uniform somatic ciliation and specialized oral cilia within a buccal cavity.

hyphae The microfilaments composing the mycelium of a fungus.

hypochlorite OCl^-, chlorine anion commonly used as an alternative to chlorine gas for disinfection.

hypotrich A subclass of ciliates in the class Spirotrichea. Typically with a dorsoventrally flattened body with cirri (compound ciliary organelles) on the ventral surface and with conspicuous oral cilia (rows of membranelles).

illuminator A device for producing, concentrating, or reflecting light.

immersion Placement into or within a fluid, usually water.

immunization Natural or artificial development of resistance to a specific disease.

immunocompromised Having the immune system impaired or weakened.

in situ In the original location.

incineration The process of reducing the volume of a solid by the burning of organic matter.

indicator species A species whose requirements most reflect those of the species community in the habitat of concern, usually used to indicate habitat quality and to predict future conditions.

inert Lacking an activity, reactivity, or effect.

influent Water or wastewater flowing into a basin or treatment plant.

ingestion The act or process of taking food and other substances into the animal or protozoan body.

inhibition The act of repressing or restraining a physical or chemical reaction.

inorganic Pertaining to or composed of chemical compounds that do not contain carbon as the principle element (excepting carbonates, cyanides, and cyanates), that is, matter other than plant or animal.

interior Of, relating to, or located in the inside; inner.

intermediate A precursor to a desired product.

intermediate host The host in which a parasite multiplies asexually.

intracellular Within a cell.

ion exchange A chemical reaction in which mobile hydrated ions of a solid are exchanged, equivalent for equivalent, for ions of like charge in solution.

ionize To convert totally or partially to ions.

iris diaphragm An adjustable diaphragm of thin opaque plates that can be turned by a ring to regulate the aperture of a lens.

ISO/ASA speed The speed rating for a particular film, for example, a film having an ISO rating of 25 is a slow film and needs a lot of exposure.

isomerization A process by which a compound is changed into an isomer, for example, conversion of butane to isobutane.

isothermal Having constant temperature.

jar test A test procedure using laboratory glassware for evaluating coagulation, flocculation, and sedimentation in a series of parallel comparisons.

juvenile Physiologically immature or undeveloped; young.

ketone One of a class of chemical compounds of the general formula RR'CO where R and R' are alkyl, aryl, or heterocyclic radicals; the groups R and R' may be the same or different or incorporated to a ring. The ketones, acetone, and methyl ethyl ketone are used as solvents, and ketones in general are important intermediates in the synthesis of organic compounds.

kineties Rows of cilia in ciliates.

kinetoplastid A class of small flagellate protozoa in the phylum Euglenozoa; some are pathogenic parasites to humans and other vertebrates.

larva The early form that hatches from the egg of many insects, alters chiefly in size while passing through several molts, and is finally transformed into a pupa or chrysalis from which the adult emerges.

leachate Fluid that percolates through solid materials or wastes and contains suspended or dissolved materials or products of the solids.

lecithin Any of a group of phospholipids having the general composition $CH_2OR_1 \cdot CHOR_2 \cdot CH_2OPO_2OHR_3$ in which R_1 and R_2 are fatty acids and R_3 is choline and having emulsifying, wetting, and antioxidant properties.

ligand The molecule, ion, or group bound to the central atom in a chelate or a coordination compound.

lignin A substance that, with cellulose, forms the woody cell walls of plants and cements them together.

lime The term usually refers to ground limestone (calcium carbonate), hydrate lime (calcium hydroxide), or burned lime (calcium oxide).

lipids One of a class of compounds that contain long-chain aliphatic hydrocarbons and their derivatives, such as fatty acids, alcohols, amines, amino alcohols, and aldehydes; includes waxes, fats, and derived compounds.

lipolytic Capable of hydrolyzing fats, oils, or waxes.

lipopolysaccharide Any of a class of conjugated polysaccharides consisting of a polysaccharide combined with a lipid.

lipoprotein Any of a class of conjugated proteins consisting of a protein combined with a lipid.

locomotion The act of moving or the ability to move from place to place.

longitudinal Pertaining to the lengthwise dimension.

loricate Having a lorica (a hard protective case or shell).

lyophilization Rapid freezing of a material, especially biological specimens for preservation, at a very low temperature, followed by rapid dehydration by sublimation in a high vacuum.

lyse (breakup) To undergo *lysis* (the rupture of a cell that results in loss of its contents).

macroinvertebrate Any larger-than-microscopic animal that has no backbone or spinal column.

macromolecule A large molecule in which there is a large number of one or several relatively simple structural units, each consisting of several atoms bonded together.

macronucleus A relatively large densely staining nucleus that is believed to exert a controlling influence over the trophic (nutritional) activities of most ciliated protozoa.

macronutrients Elements, such as potassium and nitrogen, essential in large quantities for cellular growth.

macroorganisms Any animal or plant life that is larger than microscopic.

macrophytic Pertaining to a macroscopic plant in an aquatic environment.

magnification The apparent enlargement of an object by an optical instrument.

malicious Given to, marked by, or arising from malice.

malodorous Having a bad odor; foul-smelling.

manifold A pipe fitting with several lateral outlets for connecting one pipe to another.

marsh An area of low-lying wet land.

mass spectrometry An analytical technique for identification of chemical structures, determination of mixtures, and quantitative elemental analysis, based on application of the mass spectrometer.

mastax The muscular pharynx in rotifers.

mean cell residence time (MCRT) Average time that a given unit of cell mass stays in the activated sludge biological reactor (aeration tank). It is usually calculated as the ratio of total mixed liquor suspended solids in the reactor to that of wastage.

mesophile Bacteria that grow best at temperatures between 25 and 40 °C.

mesophilic That group of bacteria that grow best within the temperature range of 25 to 40 °C.

mesosaprobic Moderately to heavily polluted.

metabolic Of or resulting from metabolism.

metabolism The physical and chemical processes by which foodstuffs are synthesized into complex elements, complex substances are transformed into simple ones, and energy is made available for use by an organism.

metabolize To transform by metabolism; to subject to metabolism.

metalization The coating of a metal or nonmetal surface with a metal, as by metal spraying or vacuum evaporation.

metaphosphate The inorganic anion PO_3^- or a compound containing it.

metazoa Micro- and macroscopic multicellular organisms with cells organized in layers or groups as specialized tissues or organ systems.

methane formers (methanogens) Group of anaerobic bacteria responsible for conversion of organic acids to methane gas and carbon dioxide.

methanogenesis The biological production of methane.

methemoglobinemia A pathological condition caused by chemical interference with the oxygen-transfer mechanism of the blood. It may be caused in infants by drinking water high in nitrates.

microaerophilic Pertaining to those microorganisms requiring free oxygen but in very low concentration for optimum growth.

microbial Pertaining to microorganisms too small to be seen with the naked eye.

microbial community All microbial organisms within an environment.

microbial ecology The study of the interactions of microorganisms with their physical environment and each other.

microcarriers See biomass carriers.

microflora Microscopic plants.

microfungus Fungus with a microscopic fruiting body (organ specialized for producing spores).

microhabitat A small, specialized, and effectively isolated location.

micrometer An instrument used with a microscope to measure minute distances.

micronucleus A minute nucleus concerned with reproductive and genetic functions in most ciliated protozoans.

micronutrients Nutrients, such as magnesium and calcium, required in minute quantities for cellular growth.

microorganism Microscopic organism, either plant or animal, invisible or barely visible to the naked eye. Examples are algae, bacteria, fungi, protozoa, and viruses.

microtubules Any of the minute cylindrical structures that are widely distributed in the protoplasm and are made up of longitudinal fibrils.

midgut The mesodermal intermediate part of an invertebrate intestine.

mineralization The conversion of an organic material to an inorganic form by microbial decomposition.

mitochondria Any of various round or long cellular organelles of most eukaryotes that are found outside the nucleus; produce energy for the cell through cellular respiration; and are rich in fats, proteins, and enzymes.

mixed liquor suspended solids (MLSS) The concentration of suspended solids in activated sludge mixed liquor, expressed in milligrams per liter.

mixotroph An organism able to assimilate organic compounds as carbon sources while using inorganic compounds as electron donors.

molt To cast off an outer covering periodically.

monomer A simple molecule that is capable of combining with a number of like or unlike molecules to form a polymer; a repeating structure unit within a polymer.

monosaccharide A carbohydrate that cannot be hydrolyzed to a simpler carbohydrate; a polyhedric alcohol having reducing properties associated with an actual or potential aldehyde or ketone group; classified on the basis of the number of carbon atoms as triose (3C), tetrose (4C), pentose (5C), and so on.

morphology The structure or form of an organism or feature of an organism.

most probable number Statistical analysis technique based on the number of positive and negative results during testing of multiple portions of equal volume.

motile Being capable of spontaneous movement.

mucous Of, relating to, or resembling mucus.

mucus A viscid slippery substance secreted by the mucous membranes that line body passages and cavities that communicate directly or indirectly with the exterior.

mutagen An agent that rises the rate of mutation above the spontaneous rate.

mutation An abrupt change in the genotype of an organism, not resulting from recombination; genetic material may undergo qualitative or quantitative alteration or rearrangement.

mutualism A necessary and beneficial interaction between two organisms living in the same environment.

mycelial Relating to the mass of interwoven filamentous hyphae that form the vegetative portion on the thallus of a fungus and are often submerged in another body (as of soil, organic matter, or the tissues of a host).

mycolic acids Saturated fatty acids found in the cell walls of mycobacteria, *Norcardia*, and corynebacteria. Chain lengths can be as great as 80, and the mycolic acids are found in waxes and glycolipids.

mycoplasma Any of the genus (*Mycoplasma* of the family Mycoplasmataceae) of pleomorphic Gram-negative chiefly nonmotile bacteria that are mostly parasitic—usually in mammals; called pleuropneumonia-like organism.

nassulids Any ciliate of the class Nassophorea.

Neisser stain A staining process used to identify bacteria.

nematode Any member of a group of unsegmented worms.

nematophagus Organisms, such as fungi, that feed on nematodes or nonsegmented roundworms.

neutrophile An organism that grows best under neutral conditions (pH of 7).

niche The role of an organism within an environment.

nitrification Formation of nitrous and nitric acids or salts by oxidation of the nitrogen in ammonia; specifically, oxidation of ammonium salts to nitrites and oxidation of nitrites to nitrates by nitrifying bacteria.

nitrogenous biochemical oxygen demand The portion of biochemical oxygen demand whereby oxygen consumption results from the oxidation of nitrogenous material; measured after the carbonaceous oxygen demand has been satisfied.

nitrosamine Yellow, aromatic, organic compound.

nocardioforms A group of actinomycetes that form mycelia that break up into rod-shaped or coccoid elements. Certain genera in this group are pathogenic to man.

***n*-octanol–water partition coefficient** The equilibrium ratio, commonly expressed as a logarithm, of the concentration of a solute in two immiscible solvents, one less polar than the other, that is, octanol and water.

nonmotile Not capable of spontaneous movement.

non-spore-forming bacteria Bacteria that do not form endospores under adverse environmental conditions.

nucleophile A species possessing one or more electron-rich sites, such as an unshared pair of electrons or the negative end of a polar bond.

nucleus A small mass of differentiated protoplasm rich in nucleoproteins and surrounded by a membrane; found in most animal and plant cells; contains chromosomes; and functions in metabolism, growth, and reproduction.

nutrient deficiency Lack of substances that are assimilated by an organism and promote growth; generally applied to nitrogen and phosphorus in wastewater but also to other essential and trace elements.

nymphs Any of various immature insects, especially a larva of an insect (as a grasshopper, true bug, or mayfly) with incomplete metamorphosis that differs from the imago, especially in size and in its incompletely developed wings and genitalia.

objective lens (objective) A lens that forms an image of an object.

obligate aerobes Bacteria that can survive only in the presence of dissolved oxygen.

obligate anaerobes Bacteria that can survive only in the absence of dissolved oxygen.

ocular micrometer A scale in the field of vision of an eyepiece used as a measuring device.

ogliosaprobic Slightly polluted.

organelle A specialized cellular part that is analogous to an organ.

organic See organic matter.

organic matter Chemical substances of animal or vegetable origin, or more correctly, containing carbon and hydrogen.
particulate—material of plant or animal origin that is suspended in fluid.
soluble—capable of being dissolved in a fluid.

organic substrate Carbon- and hydrogen-containing compounds that are amenable to biological degradation.

orthophosphate (1) A salt that contains phosphorus as $(PO_4)^{-3}$. (2) Product of hydrolysis of condensed (polymeric) phosphates. (3).A nutrient required for plant and animal growth.

osmotic Of or relating to osmosis (the passage of a pure solvent through a semipermeable membrane).

osmotroph An organism that can absorb nutrients through the cell surface, as is commonly the case with bacteria and fungi.

oxic Presence of air or oxygen

oxidant Compound that gives up oxygen easily, removes hydrogen from another compound, or attracts negative electrons. Also known as an oxidizing agent.

oxidation (1) A chemical reaction in which the oxidation number (valence) of an element increases due to the loss of one or more electrons by that element. Oxidation of one element is accompanied by simultaneous reduction of the other reactant. (2) The conversion of organic materials to simpler, more stable forms with the release of energy, accomplished by chemical or biological means. (3) The addition of oxygen to a compound.

oxidation–reduction potential The potential required to transfer electrons from an oxidant to a reductant that indicates the relative strength potential of an oxidation–reduction reaction.

oxygenase An oxidoreductase that catalyzes the indirect incorporation of oxygen to its substrate.

oxygenated Treated, infused, or combined with oxygen.

oxygenic See aerobic.

ozone O_3, a strong oxidizing agent with disinfection properties similar to chlorine; also used in odor control and sludge processing.

parasite An organism living in or on another organism.

parthenogenetic Pertaining to reproduction in which offspring are produced by an unfertilized female; usually used when the species in question normally reproduces sexually.

particulate Matter in the form of small liquid or solid particles.

partition To divide into parts, pieces, or sections.

passive transport See facilitated diffusion.

Pasteur pipette An ungraduated glass pipette, typically used to transfer small volumes of liquid sample to a microscope slide.

pasteurization The application of heat for a specified time to a liquid food or beverage to enhance its keeping properties by destroying harmful microorganisms.

pathogen A disease-producing agent; usually refers to living organisms.

pathogenic organisms Organisms that cause disease in the host organism by their parasitic growth.

pellet A small, solid, or densely packed ball or mass.

peptidoglycan A polymer that is found in bacterial cell walls.

peritonitis Inflammation of the peritoneum.

peritrich A bell-shaped or tubular microorganism of the class Peritria, characterized by a wide oral opening surrounded by cilia.

permeability The ability of a membrane or other material to permit a substance to pass through it.

permeable Having a texture that permits water to move through perceptibly.

permease Any of a group of enzymes that mediate the phenomenon of active transport.

persistence The tendency of a refractory, nonbiodegradable material to remain essentially unchanged after being introduced to the environment.

pH The reciprocal of the logarithm of the hydrogen ion concentration in gram moles per liter. On the 0 to 14 scale, a value of 7 at 25 °C (77 °F) represents a neutral condition. Decreasing values indicate increasing hydrogen ion concentration (acidity), and increasing values indicate decreasing hydrogen ion concentration (alkalinity).

phagocytosis (1) The process by which a cell is engulfed and broken down by another for purposes of defense or sustenance. (2) The uptake of extracellular materials by the formation of a pocket from the cellular membrane and its subsequent pinching off.

phagotroph An organism that ingests nutrients by phagocytosis.

pharyngeal Relating to, located, or produced in the region of the pharynx.

pharyngitis Inflammation of the pharynx.

phase-contrast microscope A microscope that translates differences in the phase of light transmitted through or reflected by the object into differences of intensity in the image.

phenol An organic pollutant known as carbolic acid occurring in industrial wastes from petroleum-processing and coal-coking operations.

phosphatase Enzyme that hydrolyzes phosphoric acid esters of carbohydrates.

phospholipids Any of a class of esters of phosphoric acid containing one or two molecules of fatty acid, an alcohol, and a nitrogenous base.

phosphorylate To change (an organic substance) into an organic phosphate.

photoautotroph An organism that derives energy from light and manufactures its own food.

photoheterotroph An organism using light as a source of energy and organic materials as a carbon source.

photolysis The use of radiant energy to produce chemical changes.

photomicroscopy Photographic microscopy.

photophosphorylation Phosphorylation induced by light energy in photosynthesis.

photoreceptor A highly specialized, light-sensitive cell or group of cells containing photopigments.

photosynthesis The process of converting carbon dioxide and water to carbohydrates, activated by sunlight in the presence of chlorophyll.

phototroph An organism that utilizes light as a source of metabolic energy.

phycocyanin Any of the bluish-green protein pigments in the cells of blue-green algae.

phycoerythrin Any of the red protein pigments in the cells of red algae.

physiochemical Pertaining to physiology and chemistry.

physiological In accord with or characteristic of the normal functioning of a living organism.

phytoplankton Plankton consisting of plants, such as algae.

phytotoxic Toxic to plants.

pilus (plural pili) Any filamentous appendage other than flagella on certain Gram-negative bacteria.

pin floc Small floc particles that agglomerate poorly.

pinpoint floc See pin floc.

plaque A clear area representing a colony of viruses on a plate culture formed by lysis of the host cell.

plasmid An extrachromosomal genetic element found in certain bacteria.

plasmolysis Shrinking of the cytoplasm away from the cell wall due to exosmosis by immersion of a plant cell in a solution of higher osmotic activity.

plug flow Conditions in which fluid and fluid particles pass through a tank and are discharged in the same sequence that they enter.

pneumatic controller A valve in which compressed air forced against a diaphragm is opposed by a spring to regulate fluid flow.

pneumonia An acute or chronic inflammation of the lungs caused by numerous microbial, immunological, physical, or chemical agents.

polaritization The process of producing a relative displacement of positive and negative bound charges in a body by applying an electric field.

polarity Quality of the electrical charge, whether it is positive or negative, on particles or electrodes.

polio An infectious viral disease occurring mainly in children and, in its acute forms, attacking the central nervous system and producing paralysis, muscular atrophy, and often deformity.

poly-ß-hydroxybutyrate A common organic carbon storage material of prokaryotic cells consisting of a polymer of beta hydroxybutyrate or other beta-alkanoic acids.

polychaete A small worm common in seas and estuaries and often chosen for bioassays of coastal regions.

polyelectrolyte A compound consisting of a chain of organic molecules used as coagulants or coagulant aids. See also polymer.

polymer (1) Substance made of giant molecules formed by the union of simple molecules (monomers). (2) Common term for polyelectrolyte.

polymerization The bonding of two or more monomers to produce a polymer.

polypeptide A chain of amino acids linked together by peptide bonds but with a lower molecular weight than a protein; obtained by synthesis or by partial hydrolysis of protein.

polyphosphate Inorganic compound in which two or more phosphorus atoms are joined together by oxygen.

polypropylene A crystalline, thermoplastic resin made by the polymerization of propylene, C_3H_6; the product is tough and hard, resists moisture, oils, and solvents, and withstands temperatures up to 170 °C; used to make molded articles, fibers, film, rope, printing plates, and toys.

polysaccharide A carbohydrate composed of many monosaccharides.

polysaprophic Very heavily polluted.

polyurethane foam A solid or spongy cellular material produced by the reaction of a polyester (such as glycerin) with a diisocyanate (such as toluene diisocyanate) while carbon dioxide is liberated by the reaction of a carboxyl with the isocyanate; used for thermal insulation, soundproofing, and padding.

population dynamics The aggregate of processes that determine the size and composition of any population.

posterior The hind end of an organism.

precipitation The process of producing a separable solid phase within a liquid medium; represents the formation of a new condensed phase, such as vapor or gas condensing to liquid droplets; a new solid phase gradually precipitates within a solid alloy as a result of slow, inner chemical reaction; in analytical chemistry, precipitation is used to separate a solid phase in an aqueous solution.

predator–prey relationship
predators—animals that prey on other animals as a source of food.
prey—animals consumed by other animals (predators).

primary producers Organisms that convert the energy of the sun to carbon-based energy.

product A substance produced by a chemical change.

prokaryotic Lacking a distinct nucleus.

proliferation The growth or production by multiplication of parts, as in budding or cell division.

propagation The process of increasing in number.

propionic acid CH_3CH_2COOH, water and alcohol soluble, clear, colorless liquid with pungent aroma; boils at 140 °C; used to manufacture various propionates in nickel-electroplating solutions for perfume esters and artificial flavors, for pharmaceuticals, and as a cellulsics solvent. Also known as methylacetic acid; propanoic acid.

prosthecate bacteria Single-celled microorganisms that differ from typical unicellular bacteria in having one or more appendages that extend from the cell surface.

prostomium The portion of the head anterior to the mouth in annelids and mollusks.

protein Any of a class of high-molecular-weight polymer compounds composed of a variety of α-amino acids joined by peptide linkages.

proteinaceous Pertaining to any material having a protein base.

proteolytic Enzymes that hydrolyze proteins.

protoplast The living portion of a cell considered as a unit; includes the cytoplasm, the nucleus, and the plasma membrane.

protozoa Predominantly single-celled, phagotrophic, eukaryotic organisms, including amoebae, ciliates, and flagellates.

pseudopodia Temporary projections of the protoplast of amoeboid cells in which cytoplasm streams actively during extension and withdrawal.

psychrophile An organism that thrives at low temperatures.

pupa An intermediate, usually quiescent, stage of a metamorphic insect (as a bee, moth, or beetle) that occurs between the larva and the imago, is usually enclosed in a cocoon or protective covering, and undergoes internal changes by which larval structures are replaced by those typical of the imago.

putrefaction Decomposition of organic matter, particularly the anaerobic breakdown of proteins by bacteria, with the production of foul-smelling compounds.

pyridine C_5H_5N, organic base; flammable, toxic yellowish liquid, with penetrating aroma and burning taste; soluble in water, alcohol, ether, benzene, and fatty oils; boils at 116 °C; used as an alcohol denaturant, solvent, in paints, medicine, and textile dyeing.

pyrolyze To break apart complex molecules into simpler units by the use of heat.

qualitative Of, pertaining to, or concerning quality.

quantitative (1) Of, relating to, or expressing in terms of quantity. (2) Of, relating to, or involving the measurement of quantity or amount.

quiescent Inactive or still, dormant.

random number tables A specially constructed table of numbers used in statistics to obtain a representative sample population, that is, each member of the population has an equal chance of being included in the sample.

recalcitrant Not responsive to treatment.

reducing power Electrons stored in reduced electron carriers such as NADH, NADPH, and $FADH_2$.

reduction Chemical reaction in which an element gains an electron.

refractory organic Organic substances that are difficult or impossible to metabolize in a biological system.

remedial action Something done to correct or improve a deficiency.

residence time The period of time that a volume of liquid remains in a reactor or system.

resin Any of a class of solid or semisolid organic products of natural or synthetic origin with no definite melting point, generally of high molecular weight; most resins are polymers.

resolution The process or capability of making distinguishable the individual parts of an object, closely adjacent optical images, or sources of light.

respiration Intake of oxygen and discharge of carbon dioxide as a result of biological oxidation.

return activated sludge (RAS) Settled activated sludge that is returned to mix with raw or primary settled wastewater.

rhizosphere Thin film aerobic region subject to the influence of plant roots and characterized by a zone of increased microbiological activity.

rosette A structure or marking resembling a rose.

rostrum A beak or beaklike projection.

rotating biological contactor (RBC) A fixed-growth biological treatment device whereby organisms are grown on circular disks mounted on a horizontal shaft that slowly rotates through wastewater.

rotifers Minute, multicelled aquatic animals.

saccharolytic Capable of breaking down sugars.

saprobic system Method of evaluating the overall degree of organic pollution on a single scale, using the presence or absence of individual species or community structures as parameters.

saprophyte An organism that lives on decaying organic matter.

saprozoic Feeding on decaying organic matter; applied to animals.

Sarcina A bacterial genus.

scavenger An organism that feeds on carrion, refuse, and similar matter.

scum Floatable materials found on the surface of primary and secondary settling tanks consisting of food wastes, grease, fats, paper, foam, and similar materials.

sedentary Permanently attached.

Sedgewick–Rafter method A method for the quantitative determination of microscopic organisms (i.e., those larger than bacteria) in water.

sedimentation The process of subsidence and decomposition of suspended matter carried by water, wastewater, or other liquids by gravity. It is usually accomplished by reducing the velocity of the liquid below the point at which it can transport the suspended material. Also called settling. May be enhanced by coagulation or flocculation.

segmented worm See annelid.

septic Putrid, rotten, foul smelling; anaerobic.

septum (plural septa) A partition or dividing wall between two cavities.

sessile Permanently attached to the substrate.

seta A slender, usually rigid bristle or hair.

settleability The tendency of suspended solids to settle.

sexual reproduction Reproduction involving the paired union of special cells from two organisms.

sheath A protective case or cover.

shock load A sudden increase in hydraulic or organic loading to a treatment plant.

shutter A camera attachment that exposes the film or plate by opening an aperture.

siliceous Of, relating to, or containing silica or a silicate.

single lens reflux A camera having a single lens that forms an image that is reflected to the viewfinder or recorded on film.

sinusitis Inflammation of a paranasal sinus.

sloughing The disattachment of accumulated biological solids from trickling filter media.

sludge volume index (SVI) The ratio of the volume in milliliters of sludge settled from a 1000-mL sample in 30 minutes to the concentration of mixed liquor in milligrams per liter multiplied by 1000.

sludge wasting Removal of solids, including biomass, from wastewater treatment processes.

solids retention time (SRT) The average time of retention of suspended solids in a biological waste treatment system, equal to the total weight of suspended solids leaving the system per unit of time (usually per day).

solubility The amount of a substance that can dissolve in a solution under a given set of conditions.

soluble Capable of being dissolved.

solvent That part of a solution that is present in the largest amount, or the compound that is normally liquid in the pure state (as for solutions of solids or gases in liquids).

sorption A general term used to encompass the processes of adsorption, absorption, desorption, ion exchange, ion exclusion, ion retardation, chemisorption, and dialysis.

sorption partition coefficient The concentration of a chemical adsorbed to the matrix divided by the concentration in solution.

species A taxonomic category ranking immediately below a genus and including closely related, morphologically similar individuals that actually or potentially interbreed.

spicule A minute, slender, pointed (usually hard) body.

spirillar Corkscrew shaped.

spirochaetes Any of an order (Spirochaetales) of slender spirally undulating bacteria, including those causing syphilis and relapsing fever.

spirotrich An order of ciliated protozoans having membranelles around the mouth and few cilia elsewhere on the body.

spontaneous Occurring without application of an external agency because of the inherent properties of an object.

sporozoa Also known as the Apicomplexa, is a phylum of Protozoa, typically producing spores during the asexual stages of the life cycle.

stabilization Maintenance at a relatively nonfluctuating level, quantity, flow, or condition.

stage micrometer A finely divided scale ruled on a microscope slide and used to calibrate the threads or lines across the field of view.

starch Any one of a group of carbohydrates or polysaccharides of the general composition $(C_6H_{10}O_5)_x$, occurring as organized or structural granules of varying size and markings in many plant cells; it hydrolyzes to several forms of dextrin and glucose; its chemical structure is not completely known, but the granules consist of concentric shells containing at least two fractions: an inner portion called amylose and an outer portion called amylopectin.

stereoscopic binocular microscope (stereomicroscope) An assembly of two microscopes into a single binocular microscope to give a stereoscopic (three-dimensional) view and a correct rather than an inverted image.

sterilization The destruction or removal of all living organisms within a system.

stirred sludge volume index (SSVI) Volume occupied by the solids after the settling period.

storage granule (1) Membrane-bound vesicles containing condensed secretory materials (often in an inactive zymogen form). (2) Granules found in plastids or in cytoplasm, assumed to be food reserves; often of glycogen or other carbohydrate polymer.

stramenopiles Classification of protists named for the tripartite tubular hairs associated with most members of the group; contains the zooflagellates, amoebae, funguslike organisms, sporozoan-like organisms, and varied forms of algae.

stratification The arrangement of a body of water into two or more horizontal layers of differing characteristics, especially densities.

streptococcal infection Infection with bacteria of the genus *Streptococcus*.

strict aerobes See obligate aerobes.

strontium A soft, malleable, ductile, metallic element.

stylet A relatively rigid elongated organ or appendage.

substrate A surface on which an organism grows or is attached.

substratum Any layer underlying the true soil.

succession A gradual process brought about by the change in the number of individuals of each species of a community and by the establishment of new species populations that may gradually replace the original inhabitants.

suctorian ciliates A subclass of ciliates in which adults are nonmotile, predatory, and unciliated and usually with suctorial tentacles for food capture.

sulfate-reducing bacteria Bacteria capable of reducing sulfate or other forms of oxidized sulfur to hydrogen sulfide.

sulfate reduction A process by which microorganisms reduce sulfate to organic sulfhydryl groups (R-SH).

sulfur oxidation Oxidation of elemental sulfur or hydrogen sulfide by microorganisms to hydrogen sulfate.

supernatant The liquid above settled solids.

surfactant A surface-active agent such as detergent that, when mixed with water, generally increases its cleaning ability, solubility, and penetration while reducing its surface tension.

suspension A mixture of fine, nonsettling particles of any solid within a liquid or gas.

symbiosis The living together of two dissimilar organisms, in which the association generally is advantageous or even necessary to one or both and is not harmful to either. Such a phenomenon is found among organisms in biological treatment processes.

synergistic An interaction between two entities producing an effect greater than a simple additive one.

synthesis Any process or reaction for building up a complex compound by the union of simpler compounds or elements.

tardigrades Any of a phylum (Tardigrada) of microscopic arthropods with four pairs of stout legs that usually live in water or damp moss. Also called water-bears.

taxonomy The study of the general principles of scientific classification.

telotroph An annelid larva having a preoral and posterior band of cilia.

teratogen An agent causing formation of a congenital anomaly.

terminal electron acceptor The final molecule that gets electrons during an oxidation–reduction reaction.

terrestrial Growing or living on land, as opposed to the aquatic environment.

testate Having or covered by a test (an external protective or skeletal covering), usually calcerous, silaceous, chitinous, fibrous, or membranous, secreted or built.

tetrad A group or arrangement of four.

thermophile Bacteria that grow best at temperatures between 45 and 60 °C.

thermophilic That group of bacteria that grow best within the temperature range of 45 to 60 °C.

thermotolerant Able to withstand high temperatures.

toxic Capable of causing an adverse effect on biological tissue following physical contact or absorption.

toxicity The property of being poisonous or causing an adverse effect on a living organism.

trace An extremely small but detectable quantity of a substance.

transmittance The fraction of radiant energy that, having entered a layer of absorbing matter, reaches its farther boundary.

treatability study A study in which a waste is subjected to a treatment process to determine whether it is amendable to treatment and to determine the treatment efficiency or optimal process conditions for treatment.

tricarboxylic acid cycle A sequence of enzymatic reactions involving oxidation of a two-carbon acetyl unit to carbon dioxide and water to produce energy for storage in the form of high-energy phosphate bonds. Also known as the Krebs cycle; citric acid cycle.

trichome A filamentous outgrowth.

trickling filter An aerobic, fixed-growth wastewater treatment process in which organic matter present in wastewater is degraded as it is distributed over a biological filter bed.

triglyceride A naturally occurring ester of normal, fatty acids and glycerol; used in the manufacture of edible oils, fats, and monoglycerides.

trinocular microscope A microscope with three ocular lenses, the third of which is used for placement of a photographic device.

trophic levels Any of the feeding levels through which the passage of energy through an ecosystem proceeds.

trophus Masticatory (chewing) apparatus in *Rotifera.*

trunk The main mass of the body, exclusive of the head, neck, and extremities; it is divided into thorax, abdomen, and pelvis.

tuberculosis A chronic infectious disease of humans and animals primarily involving the lungs, caused by the tubercle bacillus, *Mycobacterium tuberculosis*, or by *M. bovis.*

tungsten (1) Also known as wolfram. A metallic transition element, symbol W, atomic number 74, atomic weight 183.85; soluble in mixed nitric and hydrofluoric acids; melts at 3400 °C. (2) A hard, brittle, ductile, heavy gray-white metal used in the pure form chiefly for electrical purposes and with other substances in dentistry, pen points, X-ray-tube targets, phonograph needles, and high-speed tools and as a radioactive shield.

turbid Thick or opaque.

turbidity Suspended matter in water or wastewater that scatters or otherwise interferes with the passage of light through the water.

typhoid fever A highly infectious, septicemic disease of humans caused by *Salmonella typhi*, which enters the body by the oral route through ingestion of food or water contaminated by contact with fecal matter.

typhus Any of three louse-borne human diseases caused by *Rickettsia prowaszkii*, characterized by fever, stupor, headaches, and a dark-red rash.

ultraviolet radiation (UV) Light rays beyond the visible region in the visible spectrum; invisible to the human eye.

unsaturated Any chemical compound with more than one bond between adjacent atoms, usually carbon, and thus reactive toward the addition of other atoms at that point, for example, olefins, diolefins, and unsaturated fatty acids.

urea A natural product of protein metabolism found in urine; synthesized as white crystals or powder with a melting point of 132.7 °C; soluble in water, alcohol, and benzene; used as a fertilizer; in plastics, adhesives, and flameproofing agents; and in medicine. Also known as carbamide.

vacuole A membrane-bound cavity within a cell; may function in digestion, storage, secretion, or excretion.

van der Waals force An attractive force between two atoms or nonpolar molecules, which arises because a fluctuating dipole moment in one molecule induces a dipole moment in the other, and the two dipole moments then interact.

vapor pressure For a liquid or solid, the pressure of the vapor in equilibrium with the liquid or solid.

vaporize To convert a liquid to a gas.

vegetative Of, relating to, or engaged in nutritive and growth functions.

ventral On or belonging to the lower or anterior surface of an animal, that is, on the side opposite the back.

virulence The disease-producing power of a microorganism; infectiousness.

virus The smallest (10 to 300 μm in diameter) life form capable of producing infection and diseases in humans and other animals.

viscosity The degree to which a fluid resists flow under an applied force.

volatile A substance that evaporates or vaporizes at a relatively low temperature.

volatilization The conversion of a chemical substance from a liquid or solid state to a gaseous or vapor state by the application of heat, the reduction of pressure, or a combination of these processes.

volute A twisted or spiral formation.

vulva The external genital organs of females.

waterbear A microscopic arthropod with four pairs of stout legs that lives in water or damp moss.

water hyacinth Floating aquatic plants whose roots provide a habitat for a diverse culture of aquatic organisms that metabolize organics in water.

wax Any group of substances resembling beeswax in appearance and character, and in general distinguished by their composition of esters and higher alcohols and by their freedom from fatty acids.

weirs A baffle over which water flows.

wetlands Surface areas, including swamps, marshes, and bogs, that are inundated or saturated by groundwater, often enough to support a prevalence of vegetation adapted to life in saturated-soil conditions.

wet mount Preparation of a microscope slide by placing a drop of sample water directly on the slide.

worst-case A testing situation in which the most unfavorable possible combination of circumstances is evaluated.

xenobiotic A completely synthetic chemical compound that does not naturally occur on earth.

yolk gland A modified part of the ovary that, in many flatworms and rotifers, produces yolk-filled cells serving to nourish true eggs.

zone settling When particle concentrations are sufficient enough so that particles interfere with the settling of other particles and they stick together.

zoogleal A jellylike matrix developed by certain bacteria. A major part of activated sludge floc and trickling filter slimes (zooglea).

Index

A

B

C

D

E

F

G

H

I

K

L

M

N

O

P

R

S

T

U

V

W

Z